INTRODUCTION TO
PLANT NEMATOLOGY

INTRODUCTION TO PLANT NEMATOLOGY

Second edition

VICTOR H. DROPKIN
Department of Plant Pathology
University of Missouri, Columbia

WILEY

A WILEY-INTERSCIENCE PUBLICATION

JOHN WILEY & SONS

New York · Chichester · Brisbane · Toronto · Singapore

Library of Congress Cataloging-in-Publication Data

Dropkin, Victor Harry, 1916 –
    Introduction to plant nematology/Victor H. Dropkin. — 2nd ed.
        p.   cm.
    "A Wiley-Interscience publication."
    Includes bibliographies and index.
        ISBN 0-471-85268-6
    1. Plant nematodes.   2. Nematode diseases of plants. I. Title.
SB998.N4D76 1988   88-14867
632'.65182—dc19   CIP

Printed in the United States of America

10   9   8   7   6   5   4   3   2   1

To Elizabeth and Greg

Preface

The purpose of this revision is to present a contemporary view of phytonematology. Since publication of the first edition some new directions of
research have gained prominence. Computer modelling of nematode population dynamics, DNA and protein analysis to identify species and subspecific variants, and increased attention to scanning electron microscopy
are now entering the mainstream of knowledge. Advances in biological
science are changing requirements for training students in phytonematology.

The text has been completely rewritten. Discussion of each genus is
expanded to include sections on morphology, biology, pathology, interaction with other pathogens, control, and a brief review of recent research.
Thus information on each genus is assembled together. Ecology is presented in a separate chapter.

The struggle to sustain agriculture in the years ahead is intensifying as
human populations increase and as resources of land decrease. Knowledge
of phytonematology will grow in importance and agriculturists will need
to receive some training in the subject.

I thank Dr. Alfred Novak who read the entire manuscript and offered
valuable advice. Thanks are also due to the numerous authors and pub-
lishers who generously granted permission for use of figures from published
work. They are acknowledged throughout the text. Most of all I am grateful
to my wife, Elizabeth, for her steadfast support and help throughout the
writing.

No text can be permanent. I trust that the second edition of Introduction
to Plant Nematology reflects changes and increases in knowledge and is
an improved tool for teaching.

Victor H. Dropkin

Columbia, Missouri
January 1988

Contents

INTRODUCTION TO
PLANT NEMATOLOGY

One

INTRODUCTION

Take a handful of soil from almost anywhere in the world, from the tops of mountains to the depths of the seas. Extract the living organisms in some water, and among the other forms of life you will find elongate, threadlike, active animals. These are nematodes. Many of them will be just visible without magnification, but others will be seen only with a good magnifying lens or a microscope. Or catch a fish, a bird, or a mammal almost anywhere in the world, dissect out its stomach and intestines, and in most cases you will find some nematodes inside.

Nematodes (the name is derived from the Greek word for thread) are elongate, tubular organisms, somewhat spindle shaped, that move like snakes. They are aquatic and live in marine and fresh waters, in films of water within soil, and in moist tissues of other organisms. The phylum Nematoda is a large one, probably second only to insects in the diversity of species that it contains. Nematodes have been known since ancient times as parasites of human beings. However, when good microscopes became available and zoologists of the nineteenth century explored the vast array of living organisms, nematodes were neglected.

Most species are free-living. Great numbers live in the soil and along the shores of marine and fresh waters. They feed on microorganisms— bacteria, fungi, and algae. Marine species are usually larger and have more intricate bodies than those in fresh water and soil. In the soil, nematodes are part of the chain of organisms that reduce the remains of larger animals and plant tissue to their primary constituents. The nematodes, by removing a portion of those organisms that directly consume organic matter, help to keep the system operating at a rapid pace.

Nematodes are also highly successful parasites. Hippocrates, a Greek physician who wrote approximately 2400 years ago, mentioned roundworm (nematode) parasites of humans, including a large intestinal form, *Ascaris*. When Europeans were moving into the tropics in the late nineteenth and early twentieth centuries, medical practice had to deal with nematode-induced diseases. The study of roundworm parasites of humans and of

1

cattle, sheep, horses, pigs, and household pets flourished. We now know that many serious human maladies are induced by nematodes: in the intestines, hookworm and *Ascaris*; in lymph vessels, filaria; in muscles, loa loa; and in the eye, *Onchocerca*.

Nematode parasites occur in most animals large enough to accommodate them, including earthworms, insects, molluscs, and vertebrates. Most free-living species have uncomplicated life cycles, but some parasites display astonishingly complex life histories. *Ascaris*, the common roundworm of the human intestine, produces eggs that reach the soil in excrement. These eggs undergo arrested development, then survive for long periods until they are ingested and development resumes. Juveniles emerge in the intestine and the young nematodes penetrate into the circulatory system. They are carried to the lungs and are coughed up to reenter the alimentary canal, having meanwhile grown to larger size. Eventually they lodge in the intestine where they feed on blood from wounds inflicted on the intestinal wall. Mature females produce millions of eggs.

The study of nematodes in plants has a brief history. In 1743 a clergyman of scientific bent opened some malformed, shriveled grains of wheat in a drop of water and examined them with his microscope. To his amazement he saw a mass of fibers that began to twist. He wrote: "I am satisfied that they are a species of aquatic animals, and may be denominated worms, eels or serpents, which they very much resemble." Vinegar eels had been seen about 90 years before this. These are non-parasitic nematodes that feed on bacteria in vinegar. The nematode of wheat (now known as *Anguina tritici*) remained the only known parasite of plants (phytonematode) until 1857. A pathologist, Kühn, described an eelworm that caused stunting of teasel, a plant whose fruits were used to comb wool. Two years later, the cause of a serious disease of sugar beets was identified as a nematode. "Vibrios" associated with galls on cucumbers were mentioned in 1855, but the first description of root-knot nematodes came later. In the ensuing years of the nineteenth century and the first four decades of the twentieth, a handful of investigators studied nematodes. Boveri's description of chromosomes in *Ascaris* laid the foundation for modern understanding of cellular mechanisms of heredity. A few taxonomists published monographs describing free-living soil, fresh-water, and marine nematodes. Plant parasites received less attention.

In the United States, Cobb almost single-handedly publicized the existence and importance of nematodes other than parasites of animals. From his work in the Department of Agriculture, a growing interest in the subject arose, and a small group of scientists continued to demonstrate damage to plants by nematodes. A few investigators in Europe also contributed important discoveries. The big change in the subject occurred when an en-

tomologist searching for new insecticides found a practical way to control
field populations of nematodes. In 1943 Carter discovered that a by-product
of the petroleum industry (D-D, dichloropropene-dichloropropane mixture) had potential as a soil fumigant for the control of nematodes and
insects (1). This stimulated rapid growth of interest in plant parasitic nematodes. In a few years several other practical nematicides were found.
Dramatic improvement in crop yields resulting from their use caught the
attention of agriculturists all over the world. Within the next decade, serious study of phytonematodes and of their effects on plants began.

Parasitic nematodes inhabit all parts of plants, including developing
flower buds, leaves, stems, and roots, and they have a great variety of
feeding habits. Some species feed only on the outermost plant tissues,
others penetrate to deeper tissues (Fig. 1.1), and still others induce their

Fig. 1.1 Scanning electron micrograph of a nematode (juvenile of root-knot nematode) entering the root of a plant. (Photo courtesy of W. P. Wergin and R. M. Sayre.)

hosts to produce special nutrient sources upon which the parasites subsist. Damage from a few nematodes is usually slight, but large populations severely injure or kill their hosts. In addition, some nematodes reduce a plant's ability to resist fungal infection, thus compounding the damage, and others transmit pathogenic viruses among plants.

Until comparatively recently, our understanding of nematode anatomy and physiology was limited because most free-living and plant parasitic species are very small. The electron microscope has revealed an unsuspected complexity of structures within these organisms. Furthermore, modern techniques are showing that nematodes are much more diverse than anyone had previously realized. A few free-living species and a few animal parasites are now maintained in liquid cultures. This has led to investigations of biochemistry and physiology without interference from associated microorganisms. Phytonematodes are being cultivated in test tubes on plant tissues without associated microorganisms. This will also lead to a much deeper understanding of nematode physiology and biochemistry, and in particular to an understanding of the processes of disease.

Very recently, an intensive study of a nematode that feeds on bacteria in soil (*Caenorhabditis elegans*) has spawned a whole new field of investigation. With modern techniques new insights into a central problem of biology are coming into view. The problem is to determine how an organism's development is regulated by hereditary information stored in its egg, and *Caenorhabditis* is an excellent model organism. It can be cultivated on bacteria; it reproduces in a few days; it is one of the simplest multicellular organisms with a complete nervous system, sense organs, and muscles. Mutations can readily be induced. Living specimens can be stored indefinitely in liquid nitrogen at very low temperatures and be brought back to full activity. *Caenorhabditis* is hermaphroditic so that pure lines are regularly produced, but some eggs develop into males under certain conditions. Consequently, knowledge of its genetics has progressed rapidly. Already significant progress has been made toward a full description of its sensory perceptions and toward an understanding of the operation of the nervous system. The location of many genes is known and the developmental history of every cell has been delineated. Techniques of genetic engineering and of molecular biology are in use. By the induction and analysis of mutations, investigators expect to obtain a much clearer picture of the control of normal development than can be attained in more complex organisms.

Agricultural nematology is also flourishing. There are centers of work in Europe, Africa, India, North America, South America, Australia, and Japan. Nematodes are taking a recognized place among the more familiar bacteria, fungi, and insects as important pests of plants. They are also recognized as important organisms in the ecology of soils.

A full understanding of nematode-induced plant diseases requires utilization of many disciplines of biology. Soil ecology describes factors affecting nematode distribution, survival, and population cycles. Biochemistry of nematodes and plants analyzes mechanisms of disease induction. Plant physiology focuses on secondary effects of foliage and root damage. Genetics is part of breeding for resistance. It also helps us to understand the frequent appearance of new strains of pathogens capable of attacking resistant varieties. Animal behavior and its close ally, neurophysiology, are part of the study of attraction to plant tissues and of chemicals to control nematodes. In short, as knowledge of phytonematodes develops, and particularly as we seek to attain high agricultural production in the presence of nematode parasites, we need to study every aspect of the interactions between these animals and plants, using the accumulated knowledge of other branches of biology and related fields that apply to our subject.

This book presents an elementary introduction to plant parasitic nematodes and the diseases they cause.

REFERENCES

General References

Bird, A. F. 1971. The Structure of Nematodes. Academic Press, New York, 318 pp.

Chitwood, B. G. and Chitwood, M. B. 1950. Introduction to Nematology. Reprinted in 1974 by University Park Press, Baltimore, MD, 334 pp. The most comprehensive treatise in English on the relationships, classification, and characteristics of all nematodes.

Christie, J. R. 1959. Plant Nematodes, Their Bionomics and Control. Univ. Florida Agric. Exp. Sn., Gainesville. Mainly disease symptoms.

Crofton, H. D. 1966. Nematodes. Hutchinson University Library, London. A brief account of nematode anatomy and physiology.

Croll, N. A., ed. 1976. The Organization of Nematodes. Academic Press, New York, 439 pp. Fourteen chapters by recognized experts.

Croll, N. A. and Mathews, B. E. 1977. Biology of Nematodes. Wiley, New York, 201 pp. A tertiary level text on zoology, organ systems, behavior, development, life cycles, survival, biology of parasites.

Decker, H. 1969. Plant Nematodes and Their Control (Phytonematology). This is a translation (Amerind Publ. Co. PVT. Ltd. 1981, New Delhi, India) of a book originally published in German (*Phytonematologie und Bekaempfung Pflanzenparasitaerer Nematoden*), VEB Deutscher Landwirtschaftsverlag, Berlin, 1969, then translated into Russian and subsequently into English. Available from the U. S. Dept. of Commerce, National Technical Information Service, Springfield, VA 22161, 540 pp.

Filipjev, I. N. and Schuurmans Stekhoven, J. H. 1941. A Manual of Agricultural Helminthology. E.J. Brill, Leiden, The Netherlands, 878 pp. A comprehensive account of the earlier literature.

Goodey, J. B. 1963. Soil and Freshwater Nematodes. Methuen, London. Taxonomy, now out of print.

Hyman, L. 1951. The Invertebrates, Vol III. McGraw-Hill, New York, 572 pp. (see pp. 197 – 455 on nematodes).

Lee, D. L. and Atkinson, H. J. 1976. Physiology of Nematodes, 2nd ed. Macmillan, London, 215 pp.

Maggenti, A. 1981. General Nematology. Springer-Verlag, New York, 372 pp. An account of the morphology of nematodes, including both parasitic and free-living forms, well illustrated with superior graphics by the author.

Southey, J. F., ed. 1978. Plant Nematology. Ministry of Agriculture, Fisheries and Food, GD1. H. M. Stationery Office, London, 440 pp. Excellent advanced text.

Thorne, G. 1961. Principles of Nematology. McGraw-Hill, New York, 553 pp. Mainly taxonomy. Contains a good historical review and much information on important plant parasites and the diseases they cause.

Wallace, H. R. 1973. Nematode ecology and plant disease. Arnold, London. 228 pp. An effort to integrate many fields of knowledge into plant nematology; it is a collection of essays.

Zuckerman, B. M. and Rohde, R. A., eds. 1981. Plant Parasitic Nematodes, Vol. 3. Academic Press, New York, 508 pp.

Zuckerman, B. M., Mai, W. F., and Rohde, R. A. eds. 1971. Plant Parasitic Nematodes, Vols. 1 and 2. Academic Press, New York, 345 pp. and 347 pp. resp.

Journals

Annals of Applied Biology. (Published in England by Cambridge Univ. Press for the Assoc. of Appl. Biologists.) Articles on field experiments and other aspects of biology and control.

Helminthological Abstracts, Series B, Plant Nematology. Published by Commonwealth Agricultural Bureaux, Farnham House, Farnham Royal, Slough, SL2 3BN, U.K. Abstracts of the world literature.

Journal of Nematology. The official journal of the Society of Nematologists (U.S.). Principally plant parasites: taxonomy, physiology, ecology, pathology.

Nematologica. Published by E. J. Brill, Leiden, The Netherlands. An international journal of nematological research. Articles on plant parasitic and free-living nematodes. Taxonomy, physiology, diseases. The first journal established to publish in the field of plant nematology.

Nematologica Mediterranea. Published by Istituto di agraria del C.N.R., Bari, Italy.

Phytopathology. Published by the American Society of Phytopathologists. Occasional articles on plant nematology.

Plant Disease. Published by American Society of Phytopathologists. Short articles on plant diseases, including those induced by nematodes, control, symptoms.

Text Reference

1. Carter, W. 1943. A promising new soil amendment and disinfectant. Science 97:383 – 384.

Two

STRUCTURE AND FUNCTION

INTRODUCTION

Nematodes are animated, flexible, tubular animals living on moist surfaces or in liquid environments. They have all the organ systems found in more complex animals but lack a circulatory system. The evolution of nematodes

7

did not lead to any other kinds of more complex or different organisms, but it resulted in a great variety of species adapted to different environments. Marine nematodes are usually larger than those in terrestrial soils and freshwater; parasites of animals are larger still. *Ascaris*, a common intestinal parasite of swine (and humans) measures 20 × 0.5 cm or more. Most species found in soil, including parasites of plants, are 1 – 2 mm long × 1/20 mm or less wide. Figure 2.1 illustrates the morphology of a soil inhabiting nematode.

BODY WALL

The body wall encloses a fluid-filled interior that is maintained under pressure and serves as a hydroskeleton against which muscles can work. The wall consists of an outer **cuticle**, an intermediate layer, the **hypodermis**, and an inner set of longitudinal **muscles**.

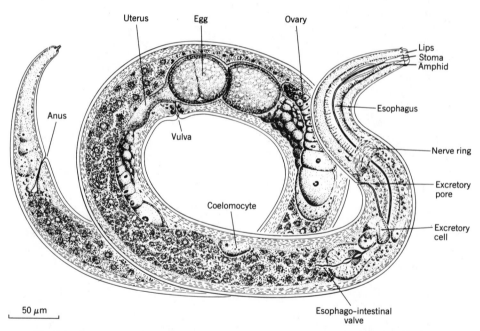

Fig. 2.1 Anatomy of an adult female free-living nematode, *Plectus parietinus*, showing location of alimentary canal (stoma, esophagus, esophago-intestinal valve, and anus), nervous system (nerve ring, sense organs—amphids), excretory system (excretory cell and pore), and reproductive system (ovary, egg, uterus, and vulva) (×280). (From A. Maggenti, General Nematology. 1981, p. 33, with permission of Springer-Verlag.)

Cuticle

The cuticle is the boundary between the nematode and the outside world. It is flexible and smooth for motion over the substrate and tough to resist wounds from abrasion. The body wall regulates water movement across the cuticle to maintain high internal pressure under changing conditions. Nematodes must also respond to stimuli from the environment, and thus their sense organs are located in the cuticle.

The cuticle of most nematodes consists of three layers: an outer cortex, a middle matrix, and an inner basal layer. Each of these may be further subdivided in certain species. The cortex, for example, often has a very thin outermost layer of lipid under which there may be several layers containing various inclusions, such as fibers. In other species, one or more layers may be absent. Apparently, each species has a cuticle adapted to its particular environment.

Although not arranged in the form of cells, the cuticle is an active structure, consisting of proteins similar to vertebrate collagen. It also contains several enzymes. Four times during the life of a nematode, the cuticle is replaced in a process called **molting** (see Fig. 2.2). At each molt the hypodermis secretes a new cuticle inside the old one and the nematode usually crawls out of the remains of the old cuticle.

As it passes through various stages of its life cycle, a nematode may encounter changed conditions of life to which it adapts by secreting a new cuticle at a molt, different in structure and function from the preceding one. In some species, a resistant juvenile stage, the dauerlarva, develops. This stage is adapted to survive starvation. Under crowded conditions, dauerlarvae appear in cultures of *Caenorhabditis*, a free- living nematode. Dauerlarvae are resistant to environmental stresses. For example, they withstand immersion in detergent solutions that kill other stages. Another example of a cuticle adapted to a particular life stage is found in cyst nematodes. The adult female of a cyst nematode (*Heterodera*) protrudes

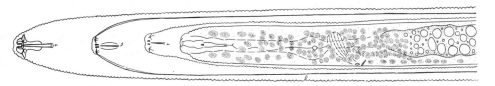

Fig. 2.2 Fourth molt of a male phytonematode (*Meloidodera floridensis*) to adult showing retained cuticles of previous molts ($\times$640). (From H. Hirschmann and A. C. Triantaphyllou. 1973. Postembryogenesis of *Meloidodera floridensis* with emphasis on the development of the male. J. Nematol. 5:18 – 195, by permission.)

from the root surface and develops a thick cuticle. The swollen body of the adult female contains several hundred embryonated eggs. When the adult dies, enzymes and substrate combine to tan the cuticle, which becomes a leathery sac that protects the enclosed eggs against drying and other stresses in the soil.

Fibers and other supporting structures are commonly present in cuticles. The fibers may be arranged in several overlapping layers. Each layer is oriented at an angle to its neighbors so that cuticle structure resembles

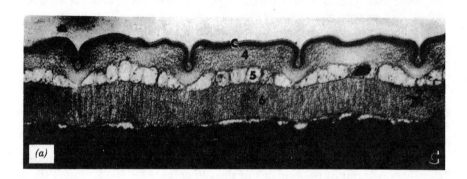

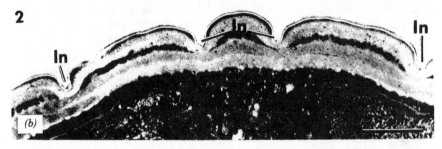

Fig. 2.3 Structure of cuticle. (a) Longitudinal section of wall of the infective stage (J–2) of *Hirschmanniella gracilis*, a phytonematode, showing four annulations and multilayered cuticle structure (×41800). (b) Cross section of lateral field of male *Heterodera* with four incisures (In) (×16400). Note expanded hypodermis in center of field. (c) Scanning electron micrograph of side view of lateral field and its anterior origin in a J–2 of *Meloidogyne hapla* (×4100). (a) (From P. W. Johnson, S. D. Van Gundy, and W. W. Thomson. 1970. Cuticle ultrastructure of *Hemicycliophora arenaria*, *Aphelenchus avenae*, *Hirschmanniella gracilis*, and *H. belli*. J. Nematol. 2:42 – 58, by permission); (b) (From J. G. Baldwin, and H. Hirschmann. 1975. Body wall fine structure of the anterior region of *Meloidogyne incognita* and *Heterodera glycines* males. J. Nematol. 7:175 – 193, by permission); (c) (From J. D. Eisenback and H. Hirschmann. 1979. Morphological comparison of second-stage juveniles of six populations of *Meloidogyne hapla* by SEM. J. Nematol. 11:5 – 16, by permission.)

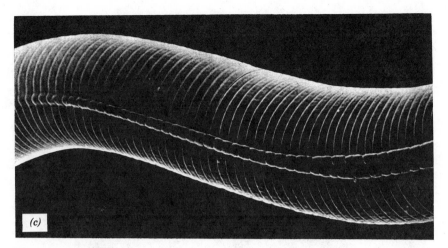

(c)

Fig. 2.3 (*Continued*)

that of a rubber tire with reinforcing fibers. This construction enables a nematode to elongate or contract while retaining the strength of its cuticle. Cuticular markings are important for distinctions among species. All nematode cuticles have surface characteristics such as longitudinal incisures, annulations (transverse rings of thinner cuticle between bands of thicker portions), various bristles, or other ornamentations. Some of these are hardly visible with standard light microscopes but are clearly demonstrated with scanning electron microscopes. Males of some species have lateral extensions of the cuticle in the tail, called **caudal alae** or copulatory **bursae**.

The formation of a highly organized noncellular cuticle that is replaced during the life cycle by a new cuticle that may differ in some details from the old one is a remarkable example of genetic control of development. We do not understand the precise mechanisms by which this is accomplished. One idea is that the cuticle may be composed of the kinds of protein that spontaneously become organized into liquid crystalline structure by virtue of their molecular properties. For example, collagen is a common protein present in tendons of humans and other organisms. It is usually present in the form of banded fibers. A solution of collagen can be prepared that exhibits no structure. But fibers with the same patterns of banding present in the original tendons will form in the solution under certain conditions. The banding patterns result from the molecular structure of the protein rather than from precise genetic control of tissue development (6). Figure 2.3 illustrates details of cuticle structure viewed through electron microscopy.

Hypodermis

The cuticle inner surface borders on a layer of tissue, the hypodermis, that secretes the cuticle. Some nematodes have a hypodermis of cells separated by membranes. In others there are no cell boundaries; the tissue is a syncytium consisting of nuclei and cytoplasm not separated by cell membranes. The hypodermis is expanded into wider portions in four locations: one lateral on each side, one dorsal, and one ventral. In the intervening portions it consists of a single layer of cells or a narrow syncytium. Major nerve trunks run longitudinally in the dorsal and ventral hypodermis. Excretory canals and nerves are contained in the lateral hypodermis.

Hypodermis has all the characteristics of an active tissue. Nuclei are

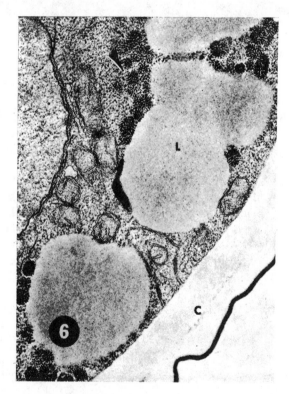

Fig. 2.4 Electron micrograph of hypodermis of *Meloidogyne incognita* 3 days after its invasion of *Impatiens balsamina*. Arrow points to glycogen. (L) lipid, (C) cuticle. Numerous mitochondria are shown (× 30,000). (From V. H. Dropkin and J. R. Acedo. 1974. An electron microscopic study of glycogen and lipid in female *Meloidogyne incognita* (root-knot nematode). Parasitol. 60:1013 – 1021, by permission.)

much larger than those of other tissues, mitochondria are numerous, and there are abundant organelles of protein synthesis (endoplasmic reticulum, ribosomes). Four times during the life of a nematode, the hypodermis produces enzymes that digest portions of the existing cuticle and secrete the new. It is also a repository of stored nutrients, including glycogen and lipids, shown in Fig. 2.4.

Muscles

The innermost lining of the body wall consists of muscles. These are composed of bands of single longitudinally elongated cells extending between their attachment at one end to the inner side of the hypodermis and at the other to the dorsal or ventral hypodermal swellings. Muscle bands thus occupy four quadrants of the nematode body, separated by the four hypodermal swellings. Each cell consists of an elongated contractile portion, a swollen belly with the cell nucleus within it, and elongated muscle arms extending to either the dorsal or ventral nerve cords. Here muscle cell extensions subdivide into several fingers that further subdivide into fine processes that make contact with the nerves. In other animals, extensions of nerves impinge upon muscles, but in nematodes muscles impinge upon nerves. Nematode muscles contract by mechanisms similar to those of other animals—proteins in thick and thin filaments slide over each other to shorten muscle fibers and cause the organism to bend (13).

NERVES

Detailed studies of nerve anatomy and physiology have focused on the larger nematodes, especially *Ascaris*. A thorough study has also been made of the anatomy of *Caenorhabditis elegans* by serial ultrathin sections examined with the electron microscope. All nematodes have a concentration of nerve cell bodies close to the midpharynx and another near the anus. Major longitudinal nerve trunks run in the ventral and dorsal hypodermal swellings and smaller nerve trunks run in the lateral hypodermal regions. The **nerve ring** is a belt of nerve connections surrounding the midpharynx. In addition, there are connections running laterally around the body between the longitudinal nerves. Sense organs, especially at the organism's anterior end, are innervated by separate nerves with cell bodies in the region of the nerve ring. The nematode nervous system has a limited number of cells (about 250 – 300) arranged to provide coordinated movements and to sense the environment.

Nerve physiology is generally similar to that of other animals. There is an acetylcholine – cholinesterase neurotransmitter system. Evidence for the action of serotonin has been obtained and there is some evidence for an inhibitory action of γ-aminobutyric acid. These findings are important for the design of antinematode compounds.

All nematodes are oriented with a lateral surface resting on the substratum. They move by coordinated contractions of the ventral and dorsal muscles. A wave of contractions passing along the dorsal side of *Ascaris* causes a parallel wave of inhibition to progress along the ventral side. This is accomplished by inhibitory nerves connecting the dorsal and ventral nerves.

Sense Organs

Nematodes have many sense organs to gather information from their environment. Each consists of one or more nerve endings (usually modified cilia) connected to a cell body at some distance. Nerve endings respond to chemical stimuli and possibly to changes in temperature or pressure. Some marine nematodes have specialized structures that are probably receptors for light. The typical sense organ is a papilla consisting of a small elevation of the cuticle within which a nerve fiber ends. Nerves also end within the cuticle without specialized structures. Two special sense organs

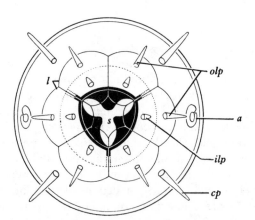

Fig. 2.5 *En face* diagram of a nematode head showing the symmetrical distribution of the amphids (a); cephalic papillae (cp); inner labial papillae (ilp); lips (l); outer labial papillae (olp); and stoma (s). (From A. F. Bird. 1971. The Structure of Nematodes. Academic Press, New York, 318 pp., by permission of Academic Press.)

are present in the anterior (**amphids**), and two in the posterior region (**phasmids**).

The arrangement of sense organs differs among species of nematodes. The basic plan, shown in Fig. 2.5, is a set of 16 + 2 sense organs on the six lips surrounding the mouth: an outer set of four cephalic papillae or setae, a middle circle of six papillae, and an inner circle of six papillae. Figure 2.6 illustrates a nerve specialized for the reception of chemical stimuli. Two specialized structures, the amphids, are present in all nematodes, located one on each side in lateral positions on the lips. As illustrated in Fig. 2.7, an amphid is an elaborate glandular and nerve structure that opens to the surface through a canal lined with cuticle.

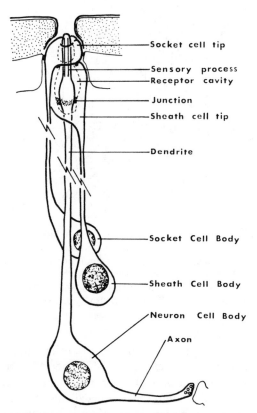

Socket cell tip

Sensory process
Receptor cavity

Junction

Sheath cell tip

Dendrite

Socket Cell Body

Sheath Cell Body

Neuron Cell Body

Axon

Fig. 2.6 Diagram of the anatomy of a nematode's sense organ for the detection of chemical stimuli. (From K. A. Wright. 1983. Nematode chemosensilla: form and function. J. Nematol. 15:151 – 158, by permission.)

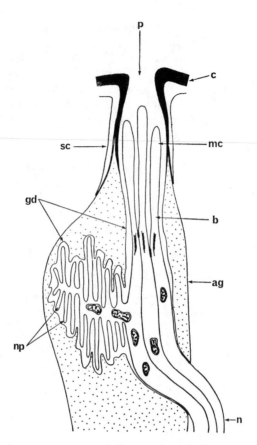

Fig. 2.7 Diagram of the relationship between the amphidial sense organ and gland in the second-stage juvenile of *Meloidogyne incognita.* (ag) Amphidial gland; (b) basal region of cilium; (c) cuticle; (gd) gland duct; (mc) modified cilium; (n) nerve axon; (np) nerve processes; (p) pore; (sc) supporting cell. (From W. P. Wergin and B. Y. Endo. 1974. Ultrastructure of the cephalic sensory organs in the southern root-knot nematode *Meloidogyne incognita.* Proc. Int. Congr. Parasitol., 3d, 1974 Vol. 1:444 – 446, by permission.)

Secretions may often be seen emanating from the amphids. Many sense organs have one or more cilia at the end of the nerves. These are probably receptors that convert environmental stimuli into nerve impulses. In many plant parasitic species the number of papillae is reduced from the basic 16.

Posteriorly there may be an aggregation of papillae in the vicinity of the anus. Nematodes of one large taxonomic group have two lateral structures, the phasmids, one on each side. These contain both glandular and nerve components. They resemble the amphids but are much smaller and probably have sensory function. Presence or absence of phasmids is used in identification, but in many species they are hard to see and pose difficulties to the novice (cf. Fig. 6.13**b**).

ALIMENTARY CANAL

The nematode alimentary canal is a tube extending from the mouth to the anus. It consists of four parts: stoma (mouth), pharynx (also called esophagus), intestine, and anus. The **stoma** has many configurations in nematodes, each adapted to the source of food. Feeders on bacteria have a cylindrical, barrel-shaped, or conical chamber that is permanently open anteriorly and connected posteriorly to the pharynx. Some species rip into tissues with extensions of the lips to get at bacteria. Carnivorous species are equipped with one or more teeth in the stoma. Some parasites of animals have an array of hooks surrounding the stoma that attach to host cells. Species that pierce plant or fungal cells use a movable, hollow or grooved spear, the **stylet**, to penetrate cell walls and to imbibe cell contents.

The **pharynx** (also called esophagus) is a muscular organ located between stoma and intestine that develops suction for ingestion of fluid. It may also contain glands in its walls, or have glands attached to it with ducts leading into its lumen. The lumen is lined with cuticle upon which radial muscles are inserted. Muscle contractions pull the walls apart, thus creating a suction. The pharynx of many nematodes is subdivided into four regions:

Procorpus—anterior portion.

Metacorpus—a wider portion behind the procorpus, sometimes in the form of a muscular pump called the median bulb.

Isthmus—a narrow section connecting the metacorpus to the basal portion.

Basal bulb—the region just anterior to the intestine, usually containing esophageal glands.

The pharynx may also have a different organization. In some species the pharynx has a uniform diameter, without distinct bulbs. Others have the procorpus and metacorpus fused into a single unit, and so on. By coordinated contractions that pass posteriorly, the pharynx moves fluid with its nutrient content into the intestine. A valve at the junction with the intestine prevents backflow into the pharynx. Species that feed on bacteria have an expanded portion at the rear of the pharynx equipped with three interlocking comblike structures to macerate bacteria.

The glandular portion of the pharynx is usually posterior and has three or more cells connected to the lumen by ducts. In many species of phytonematodes, the duct from the dorsal gland extends forward in the wall of the pharynx and empties into the lumen just behind the base of the stylet, whereas the two subventral glands empty into the midportion. In some species, however, all the gland ducts reach the pharyngeal lumen in its midregion. We may assume that each type of gland has its own particular kind of secretion. Perhaps those reaching the anterior part of the esophagus are injected into plant cells before ingestion whereas other gland secretions are mixed with esophageal contents to continue digestion en route to the intestine. Figure 2.8 illustrates the anatomy of the pharynx of a phytonematode and Fig. 2.9 is a diagram of the structure of a stylet.

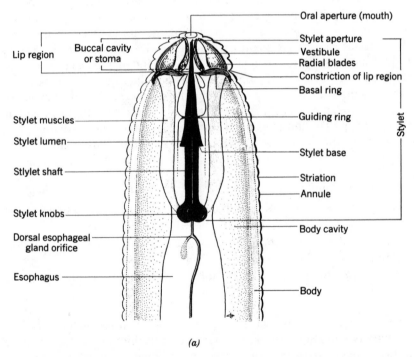

(a)

Fig. 2.8 Diagrams of anterior portion of alimentary canal. (**a**) Head region of a phytonematode; (**b**) anterior pharynx and intestine. (From S. M. Ayoub. 1977. Plant Nematology, An Agricultural Training Aid, with illustrations and photographs by C. S. Papp. Published by Dept. of Food and Agriculture, Sacramento, California, pp. 1 – 157, by permission.)

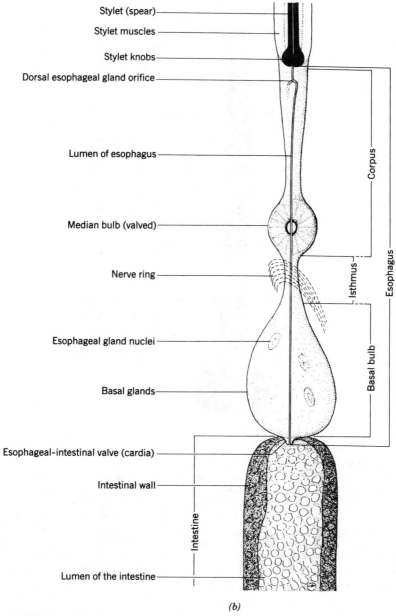

Stylet (spear)

Stylet muscles

Stylet knobs

Dorsal esophageal gland orifice

Lumen of esophagus

Median bulb (valved)

Nerve ring

Esophageal gland nuclei

Basal glands

Esophageal-intestinal valve (cardia)

Intestinal wall

Lumen of the intestine

Corpus

Isthmus

Basal bulb

Esophagus

Intestine

(b)

Fig. 2.8 (*Continued*)

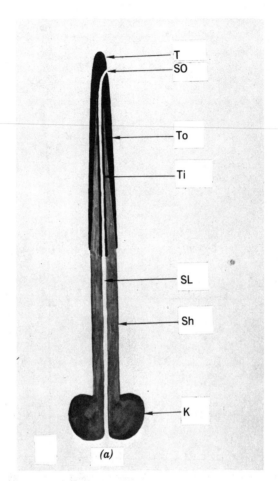

(a)

Fig. 2.9 Stylets. (a) Stylet of *Pratylenchus penetrans*, showing the interlocking tooth (T) and shaft (Sh). The tooth covers the outer portion of the anterior half of the stylet (To) as well as forming its inner lining (Ti). (SL) stylet lumen; (SO) stylet opening; (K) knob. (b) Head of *Nothocriconema annuliferus* female with protruding stylet (×2450). (a) (From T. C. Chen and G-Y. Wen. 1972. Ultrastructure of the feeding apparatus of *Pratylenchus penetrans*. J. Nematol. 4:155 – 161, by permission.); (b) (From A. T. De Grisse. 1977. De Ultrastruktuur van het Zenuwstelsel in de Kop van 22 Soorten Plantenparasitaire Nematoden, behorende tot 19 Genera (Nematoda: Tylenchida). Fakulteit van de Landbouwwetenschappen, Laboratorium voor Dierkunde, Rijksuniversiteit Gent, Belgium, 420 pp., by permission.)

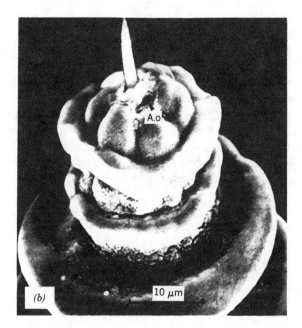

Fig. 2.9 (*Continued*)

The **intestine** is composed of large cells that may have fingerlike projections, microvilli, facing the lumen to provide surfaces for absorption. It functions both to metabolize food and to store reserves. In the light microscope the intestine appears dark because of its dense cell contents. The **anus** is held closed by the high internal pressure of the nematode's body. Muscles from anus to body wall pull the anus open during defecation.

EXCRETORY SYSTEM

There are two major kinds of excretory systems in nematodes. In some groups, a large cell in the body cavity connects to the exterior by a duct in the pharyngeal region. In others, tubes run longitudinally along the body and join together anteriorly in a duct leading to the exterior in the pharyngeal region or farther forward. The longitudinal tubes collect fluid from smaller channels that drain into the main channel. In some species the canal draining to the exterior pulsates, and the rate of pulsation slows down in hypertonic and speeds up in hypotonic media. Nematodes may also regulate ionic content by controlling movement of ions across the cuticle and across the intestine into the lumen of the alimentary canal. When nematodes are incubated in liquid, a number of organic molecules and ions can be identified as nematode products. But details of the physiology of excretion by these organisms are almost unknown.

The alimentary canal and reproductive system are tubes suspended in the body cavity, a continuous cylindrical space within the body bounded by the wall with its inner layer of muscles, intermediate hypodermis, and outer cuticle. The body cavity is filled with fluid, mostly water but also containing nutrients and other components. A nematode must maintain the composition of this fluid within limits tolerated by the cells that it bathes. Nematodes adapt to environmental stresses in ways that preserve the integrity of their internal structures. Limited evidence indicates that the cuticle is highly permeable to water and to some other compounds. Some nematodes can defend themselves against changes in ionic concentrations of the medium by eliminating excess ions.

REPRODUCTIVE SYSTEM

In most nematodes the sexes are separate and mating is required for reproduction. But in some, both sexes are combined in one individual, a hermaphrodite, in which both eggs and sperm are elaborated in the same gonad. In others, males are not required and eggs are produced without fertilization. Some species, for example, *Caenorhabditis elegans*, currently under intensive investigation, maintain populations of mostly hermaphrodites but with a small proportion of males. These mate with the hermaphrodites and the offspring consist of both males and females.

The typical female genital tract of phytonematodes is an elongated structure consisting of a distal **ovary**, followed by an **oviduct** and a **spermatheca** that stores sperm from males, and a **uterus** where fertilized eggs remain until oviposition through the **vulva** takes place. There may be one or two complete gonads in a single nematode, and the ovaries may be reflexed to accommodate their length. Some nematode females become globular in shape, the body cavity being filled with reproductive organs. A few have a prolapsed uterus that continues to grow outside the body. Some nematodes retain their eggs in the uterus until the developing embryos hatch and are then born as active juveniles. Figure 2.10 is a diagram of the reproductive tract of a nematode with two complete gonads.

The male reproductive system is also subdivided into distinct regions. The **testis** at the distal end has a germinal and a growth zone where sperm are produced and then enlarge. The **seminal vesicle** is a swollen portion in which sperm accumulate. Following this is a conducting tube, the **vas deferens**, usually glandular and muscular (Fig. 2.11). Sperm pass to the

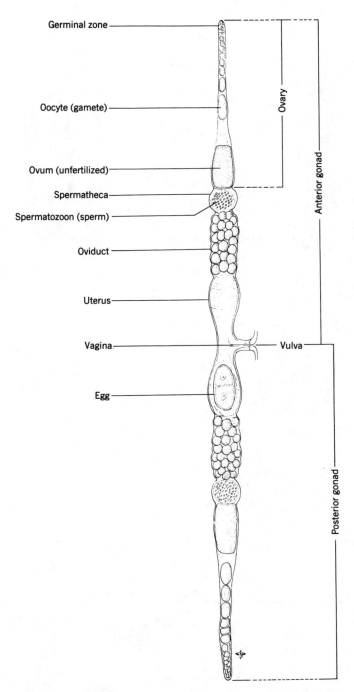

Fig. 2.10 Diagram of a female reproductive system, with two complete sets of structures. (From S. M. Ayoub. 1977. Plant Nematology, An Agricultural Training Aid, with illustrations and photographs by C. S. Papp. Published by Dept. of Food and Agriculture, Sacramento, California, pp. 1 – 157, by permission.)

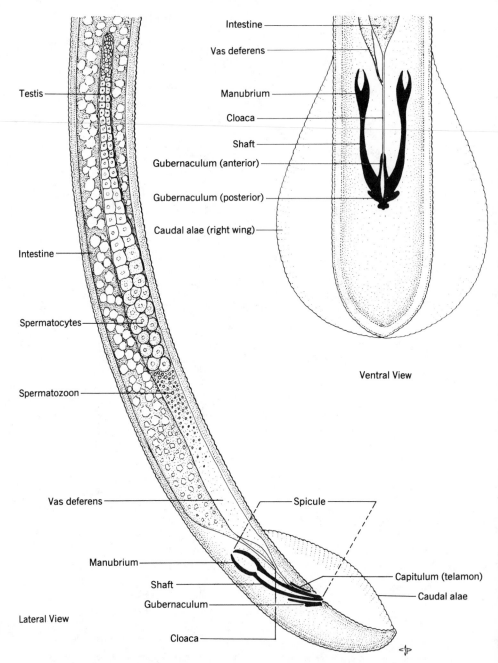

Fig. 2.11 Diagram of male reproductive system. (From S. M. Ayoub. 1977. Plant Nematology, An Agricultural Training Aid, with illustrations and photographs by C. S. Papp. Published by Dept. of Food and Agriculture, Sacramento, California, pp. 1 – 157, by permission.)

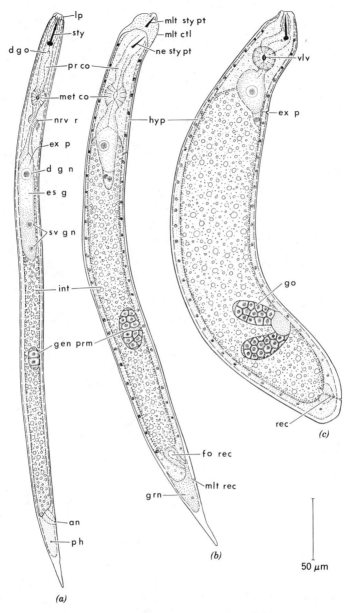

Fig. 2.12 Developmental stages of *Heterodera lespedezae*. (a) second-stage juvenile; (b) second molt; (c) third-stage juvenile; (d) third molt; (e) fourth-stage juvenile; (f) fourth molt; (g) adult female. Abbreviations: (an) anus; (ctl) cuticle; (d g n) dorsal gland nucleus; (d g o) dorsal gland orifice; (es g) esophageal gland; (ex p) excretory pore; (fo rec) forming rectum; (fo vag) forming vagina; (gen prm) genital primordium; (go) gonad; (go du) gonoduct; (grn) granules; (hyp) hypodermis; (int) intestine; (lp) lip or lip region; (lu) lumen; (met co) metacorpus of esophagus; (mlt ctl) molted cuticle; (mlt rec) molted rectum; (mlt sty pt) molted conical part of stylet; (n) nucleus; (ne ctl) new cuticle; (ne sty pt) new conical part of stylet; (nrv r) nerve ring; (ocy) oocyte;

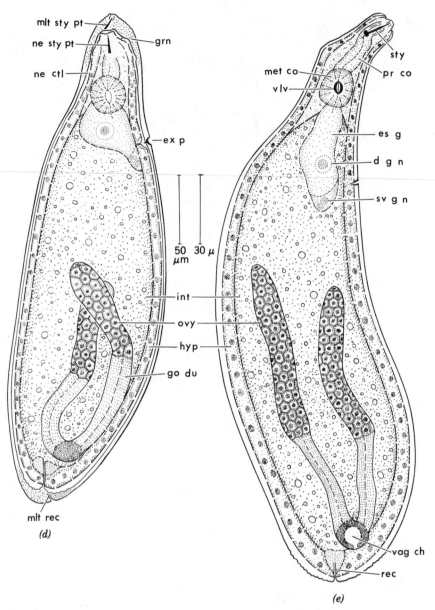

Fig. 2.12 (*Cont.*) (ovy) ovary; (ph) phasmid; (pr co) procorpus of esophagus; (rec) rectum; (sty) stylet; (sty kn) stylet knobs; (sv g n) subventral gland nucleus; (ut) uterus; (vag) vagina; (vag ch) vaginal chamber; (vlv) valve of metacorpus; (vu) vulva. (From D. S. Bhatti, H. Hirschmann, and J. N. Sasser. 1972. Post-infection development of *Heterodera lespedezae*. J. Nematol. 4:104 – 112, by permission.)

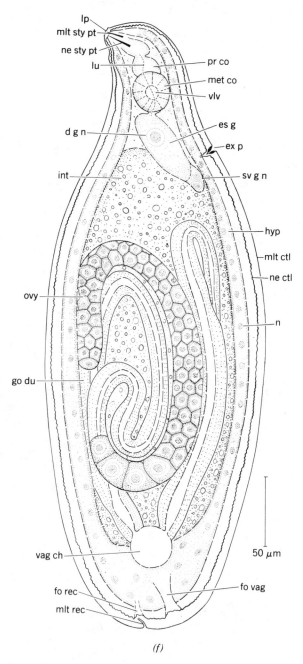

(f)

Fig. 2.12 (*Continued*)

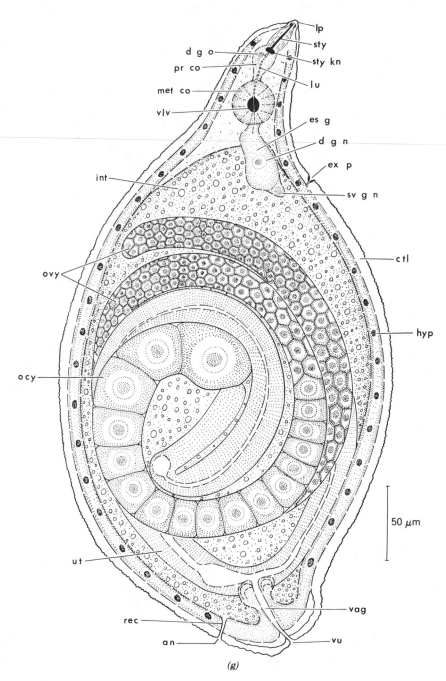

Fig. 2.12 (*Continued*)

exterior through a common opening of the reproductive system and the anus, the **cloaca**. The male gonad terminates in two hooklike structures, **spicules**, located just inside the cloaca. Together they form a passageway for sperm. During mating the spicules are inserted into the vagina of the female. Accessory glands, muscles, sense organs, and nerves are associated with the spicules. As mentioned before, males of some species have lateral cuticular extensions, the **caudal alae**, in the tail region. These are also called **bursae**.

Development proceeds by cell divisions and differentiation within the egg and subsequently during growth and maturation of juvenile stages. Many phytonematodes undergo a first molt within the egg so that emerging infective juveniles are in the second stage. Figure 2.12 shows developmental stages of a cyst nematode.

SURVIVAL

Both free-living and parasitic nematodes face periods of stress from rapid changes in their environment. The soil may lose moisture to the point where motion is restricted by the loss of films of water on soil surfaces. Heavy rainfall may flood the soil and cause oxygen in the soil water to be depleted by the metabolism of microorganisms. Temperatures may drop to below freezing or rise to high levels. Plant hosts in temperate climates die over winter, or crop rotations may remove suitable hosts. Nematodes have several mechanisms to survive environmental stresses.

Desiccation

Nematodes are mostly water (water content is 75% or more). We may therefore expect that they will lose water when exposed to dry conditions. If this loss alters the nematode's structure beyond a certain point, the organism will be irreversibly damaged and die. However, if the internal structure can be maintained, rehydration may permit life to go on. Some species display remarkable resistance to drying and survive for years under very dry conditions. For example, galls containing *Anguina tritici* were kept in a jar on a shelf in the dry atmosphere of Utah (United States) for many years. Periodically, some galls were removed, soaked in water, and examined. Nematodes revived after 32 years in storage!

In recent years we have learned that many species can survive gradual but not rapid drying. Those stylet-bearing nematodes that feed on fungi are easily maintained on agar that supports growth of a fungus. When the agar is permitted to dry, the nematodes coil themselves into tight spirals and decrease in overall length but maintain their normal width. Under conditions of slow drying, when the soil retains enough moisture to maintain more than 97% R.H. for periods of 2 weeks or more, many nematodes adapt to the dry conditions and can then survive even when they are subjected to 0% R.H. for 24 hr. The tight spiral reduces the amount of body surface exposed to evaporation. The cuticle apparently becomes less permeable to water as it dries. Under slow drying, the internal organs of the nematodes, and especially the membranes, retain their integrity. In addition to these physical adjustments to loss of water, some nematodes alter their metabolism. They may utilize their stored lipid and glycogen to synthesize protective compounds: glycerol, trehalose, or myoinositol. Thus the survival value of slow drying may be that it affords sufficient time for the nematodes to synthesize compounds that retard complete loss of water. Tightly coiled nematodes that have adapted to survive in perfectly dry conditions may have only 2 – 5% water content compared to the 75 + % of fully hydrated organisms. During drying some species aggregate into clumps. For example, *Ditylenchus dipsaci*, the bulb and stem nematode, forms masses of dauerlarvae at the base of an infected narcissus bulb. The outermost nematodes die and form a coating around the rest of the individuals, thus retarding drying. This aggregation is called "nematode wool." Rehydration also must take place slowly to permit survival of dehydrated individuals.

Some species of nematodes can withstand rapid dehydration without synthesizing trehalose or glycerol. *Plectus* spp. live in moss where frequent rapid drying occurs. These nematodes survive 0% R.H. even in the absence of an induction period. Some specimens, but not all, coil in laboratory trials, and the coiling does not contribute to survival. They survive loss of at least 90% of the internal water realized within 1 min. Obviously nematodes of this type have a different system of responding to dehydration than those that require slow drying.

The ability of nematodes to survive desiccation has some important consequences. When soils dry they may become airborne. Dust storms carry live nematodes for great distances and may deposit them in favorable locations to colonize new areas. Traps placed 2 – 3 m above the soil surface catch living nematodes of many species. Another consequence of the ability to survive under dry conditions is that dry fallow is not as effective as moist fallow in reducing nematode populations. In moist soils without host plants, nematodes remain active and consume their reserves, but in dry soils, they become quiescent and retain their food reserves (10). Quiescent nematodes, such as the tightly coiled forms described above, can resist other stresses better than active nematodes. Thus the lethal dose of nematicides required to kill dry specimens is much greater than the lethal dose for active individuals. Resistance of dehydrated nematodes to freezing and to high temperatures is also greater.

As with most other attributes, each species has its own characteristics. Limits of ability to survive environmental stress vary from one species to the next.

Freezing

Exposure to temperatures below the freezing point of water is a yearly problem for nematodes in temperate climates. If you chop out a block of frozen soil from some snow-covered grassland and bring it into a warm room, as the soil thaws, nematodes become active and appear perfectly normal. However, not all species survive freezing equally well. The danger to a nematode is twofold: first, ice crystals that form within cells can destroy membranes and kill the cells; second, free water becomes unavailable during freezing so that the nematode may become desiccated. Some species apparently tolerate cold temperatures that cause others to freeze. They may enter a supercooled state in which water is bound to organic compounds so that ice crystals do not form until the temperature descends below the temperatures that kill other species. As in the case of adaptation to desiccation, nematodes may produce antifreeze compounds such as glycerol. Because nematodes do not have a mechanism to maintain body temperatures their metabolic activity is reduced at low temperatures and under some very low temperatures metabolism appears to shut down completely. The lipid composition of certain cold-resistant species is different from that of nonresistant species. The former have lipids that maintain their structure at lower temperatures than the latter. This protects membranes from damage.

Starvation

Plant parasitic nematodes have a much higher percentage of lipid than free-living forms. This is deposited in the form of droplets in the intestinal cells. During starvation, when newly hatched juveniles fail to find suitable hosts, these lipid reserves are depleted. Some parasitic species survive periods without hosts by arresting their development at the fully formed second-stage juvenile (J–2) in the egg. These eggs with their enclosed J–2's remain viable for long periods, years in some cases. When a stimulus from a growing root of a suitable host is present, the enclosed juveniles become active, emerge from the eggs, and are attracted to growing roots.

Anoxia

Oxygen concentration in soil fluctuates as soil moisture changes. Atmospheric oxygen diffuses slowly in water and microorganisms quickly deplete the supply. When fields have standing water in them for long periods of time, what happens to the nematodes in the soil?

Knowledge of the response of plant parasitic nematodes to fluctuating oxygen levels is scanty. Individuals of many species lose their ability to move in water of low oxygen tensions. But some are adapted to survive and carry on their normal activities during periods of no or low oxygen supply. *Aphelenchus avenae*, a common soil nematode that feeds on fungi, uses lipid for energy when oxygen is abundant and changes to metabolism of carbohydrate under anoxic conditions. It can survive lack of oxygen for weeks. Other species survive anoxia by reducing their metabolism to very low levels. And some species, especially those of free-living nematodes that feed on bacteria, die when oxygen levels fall below a threshold.

BEHAVIOR

Although nematodes have seemingly simple equipment and at first glance may appear to have a limited repertoire of behavior, they are capable of complex actions. Movements possible for nematodes are:

1. Head movements in three dimensions in an irregular pattern;
2. Dorsoventral waves that move backward along the animal, propelling it forward;
3. Reversed waves, moving forward along the body, propelling it backward;
4. Deep bends in which the head and tail are brought close together;
5. Tail kinks, sharp bends near the tail; and
6. Coiling into tight spirals.

In general, phytonematodes (plant parasites) move more slowly than free-living species (2).

The life cycle of some species parasitic on animals may be intricate. For example, hookworm adults live in the human intestine attached to the wall from which they suck blood. Eggs are excreted in the feces and the juveniles that hatch climb to the top of a grass blade, where they remain until coming into contact with a foot. The nematodes then penetrate through the skin, enter the circulation, and reach the lungs, where they leave the bloodstream to enter the lung air passages. They are then coughed into the mouth, are swallowed, and ultimately reach the intestine where they attach to the epithelium to feed on blood. Other nematode parasites of animals have even more complex cycles that involve passage through invertebrate hosts. Hazards of existence are great, and nematodes produce many eggs, ensuring survival of the species. We suppose that these organisms negotiate their life cycles by responding to a set of stimuli that change as each new environment is encountered. At several points the host contributes an essential unit of behavior.

In comparison with this, parasites of plants and free-living nematodes have simple life cycles. Their behavior must respond to changing conditions of the environment and enable them to find food and mates.

The free-living species, *Caenorhabditis elegans*, is attracted to many substances, including cyclic nucleotides; anions such as Cl^-; cations, for example, Na^+, K^+, and certain amino acids (5). Plant parasitic nematodes are attracted to roots, probably by CO_2 and other substances. Some species enter roots at preferred places, such as at the tip, and move to certain locations before settling down to feed. Obviously, a nematode is well equipped with sense organs and can coordinate its movements to reach its goal.

Attraction between the sexes also operates. When a female cyst nematode is placed in one position on an agar surface it releases attractive compound(s) that diffuse out, forming a gradient. A male that comes into contact with this moves in an irregular spiral, pausing occasionally to turn in several directions, then continues and eventually finds the female. In some cases her attractive pheromone is specific so that males of foreign species are not attracted.

In addition to chemical cues, there are other stimuli to which nematodes respond.

Very slightly elevated temperatures (fractions of a degree Celsius) are attractive to certain nematodes. Higher temperatures may be avoided.

Tactile stimuli also may operate. Nematodes move along the surface of a root, probing it at intervals with the stylet until, perhaps, the right texture or the proper resistance is encountered. The individual braces itself and thrusts repeatedly until it makes a hole in an epidermal cell and enters the root.

Electrical fields affect behavior. Some species are attracted to the anode, others to the cathode. The importance of electrical fields in soil is not known, although the electrical potentials present at root surfaces are of the magnitude that attracts nematodes in experiments. Furthermore, the potential varies along the root.

Stimulus from light may also evoke a response from nematodes, but very little knowledge on this point exists.

One gets the impression from watching nematodes that they know what they are doing, although they often appear to be rather indirect in their approach to attractants and many die en route to their goal. Wallace (11) presented an excellent summary of the knowledge of phytonematode behavior current at the time of writing. Doncaster and associates at Rothamsted, England, and Wyss and associates at Kiel and Göttingen, West Germany, have produced remarkable ciné films from which they analyzed nematode behavior. Some of these are available from these authors (3,12).

METABOLISM

Nematodes maintain energy stores in the form of lipids and carbohydrates. Lipids constitute from 23 to 40% of the total dry weight, about 5 – 10 times more than that of nematode parasites of animals. Detailed analyses of lipid composition of several phytonematodes are available (8).

Carbohydrates account for about 10% of the dry weight of phytonematodes, less of free-living nematodes, and more of animal parasites. Glycogen is the principal storage carbohydrate of nematodes, with small amounts of glucose and trehalose in addition. Glycogen is stored in the hypodermis, the swollen portions of muscle cells, the intestine, and the epithelial cells of the reproductive system. Some nematodes utilize glycogen for energy during starvation in environments of low oxygen supply, then resynthesize it from lipids when oxygen is restored. During aerobic starvation, lipid reserves are metabolized. In at least one species of phytonematode, the relative amounts of lipid and carbohydrate change during development. The infective stage of *Meloidogyne* is packed with large amounts of lipid and relatively little glycogen. Within 2 – 3 days after entry into plant roots, lipid is reduced and glycogen accumulates. As adults develop, lipid reappears in great quantities and eggs are rich in both types of energy reserve (4). Nematodes live in a great variety of environments; consequently their management of lipid and carbohydrate reserves differs in adaptation to the various conditions.

Protein: In addition to structural proteins in tissues, nematodes deposit collagen in the cuticle and egg shells. Cuticles differ among nematodes, in both form and function. Adult females of the genus *Heterodera* have thick cuticles. When the gravid female dies, polyphenol oxidase is released within the cuticle, proteins are bound in a tanning reaction, and a tough wall is produced that protects eggs within the body of the female (now called a cyst). Other nematodes do not show such changes.

All species of nematodes so far studied excrete ammonia or its ion as the major component of nonprotein nitrogen. This highly toxic substance is diluted in the water of the medium. Other nitrogenous compounds are released from phytonematodes, including amino acids and amides. *Ditylenchus triformis*, a species that feeds on fungi, excretes nitrogenous compounds as ammonia (39%), amino acids (28%), volatile nitrogen other than ammonia (13%), and other nitrogen (20%). Excretions from phytonematodes in plant tissues may play a role in inducing the damage to plant tissues, but there is little evidence on this point (9).

Details of the metabolism of phytonematodes are scanty. Several enzymes of carbohydrate metabolism have been reported, suggesting that the Embden — Meyerhoff pathway of glycogen utilization operates and that the TCA cycle for production of ATP is present, but there is no definitive demonstration of complete cycles. Enzymes of the hexose monophosphate shunt have been found in a *Ditylenchus*. Information on lipid and protein metabolism is almost totally lacking. Knowledge of these systems would probably be most useful to find enzymes in nematodes that could be blocked without damaging their hosts, thus providing an additional key to control. Bolla (1) published a general review of nematode energy metabolism.

REPRODUCTIVE PHYSIOLOGY

Sexes are separate in most species. Males usually have one gonad, but in some species it is double. The testis is often reflexed to accommodate its length. Sperm are acorn shaped or they may have a tail. They move like an ameba. Females are generally larger than males. The gonad may be single or double, with the two branches joining close to the external opening. An expanded portion of the gonad, the spermatheca, is a sperm storage site and may be located at various positions, such as at the junction of ovary and oviduct or between oviduct and uterus. The uterus holds eggs until they are deposited by muscular action of the vagina or of the body as a whole. In a few species, the female body is greatly expanded to hold a very large gonad. In one species of insect parasite, the uterus prolapses out of the body and grows to enormous size, leaving the rest of the body as a small appendage.

Fertilization takes place as oocytes pass through the oviduct. Hermaphroditic females usually produce sperm first and store them in the spermatheca, then oocytes are formed. In parthenogenetic species no sperm are produced and the diploid number of chromosomes is maintained.

A typical nematode eggshell has three layers: an inner lipid layer, a relatively thick intermediate layer containing chitin and/or protein, and a thin outer layer. These layers are all secreted by the egg itself after fertilization in the oviduct. In addition, the uterus may secrete an additional thin outer layer. The fertilized embryo within the egg develops into a coiled juvenile. In most species embryonated eggs are deposited before juveniles have hatched, but in some species hatching occurs in the uterus and the female delivers motile young. Eggs are adapted to survive in the environment until conditions become favorable for the newly hatched juveniles to continue their life cycle. Parasitic species have mechanisms to endure environmental stresses such as temperature extremes or dry conditions. They respond to those conditions that are associated with the proximity of a host. For example, nematodes of many species of the genus *Heterodera* are stimulated to hatch by exudates from roots of host plants growing nearby. Eggs of some animal parasites hatch only when precise combinations of temperature, pH, and contents of the medium are present. These occur in the stomachs of their hosts! A few species of phytonematodes excrete mucoid material into which they deposit eggs. The egg sacs protect their clutch of eggs from drying.

The fertilized egg contains a single cell with dense cytoplasm containing stores of lipid and glycogen. By repeated divisions and cell differentiation, the single cell develops into an elongate juvenile. Many plant parasites molt once within the egg so that a second-stage juvenile emerges from the egg. During hatching, stylet-bearing species may cause the egg wall to soften, probably by the combined activity of the nematode and enzymes it secretes. Some species cut a precise slit at one end of the egg and then crawl out to the exterior.

NEUROMUSCULAR PHYSIOLOGY

The nervous system of nematodes can be divided into three interconnected parts: cephalic ganglia surrounding the nerve ring; longitudinal nerve cords; and caudal ganglia. Males have more nerve cells in the tail region than females. This is related to the sensory function and use of the tail during mating. Motor neurons have their cell bodies and nuclei in the ventral cord. Branches go to the dorsal cord. Axones, nerve processes that transmit impulses away from cell bodies, make contact with other nerves. Dendrites, which pass impulses toward cell bodies, also connect with other nerves. Thus the nervous system consists of an interconnected network of cells that communicate with each other. Sensory nerves receive impulses from specialized sense organs located mainly in anterior parts of the organism but also along the body. Impulses also pass directly at a slow rate between muscle cells. Speed of passage of nerve impulses is more rapid than this, but less than the speed in mammalian nerves. The anatomy of muscle-nerve junction is illustrated in Fig. 2.13.

Excitatory nerves are connected to inhibitory nerves in the same region of the body so that when muscles on the dorsal side contract, those on the ventral side relax. Thus bending movements are possible. Although depth of knowledge of nerve structure and function in *Ascaris* is considerable, we do not yet know how a nematode coordinates movements into behavior patterns (7).

In the adult *Ascaris*, somatic tissues have large numbers of cells, but the nervous system remains at 250 cells. In *Caenorhabditis* there are 300 nerve cells. In general, detailed structure and electrical properties of nematode muscles resemble those of mammals with minor differences. They are stimulated by acetylcholine at muscle-nerve junctions of stimulatory nerves. Other neurotransmitters such as serotonin, and possibly γ-aminobutyric acid, are active in nematodes.

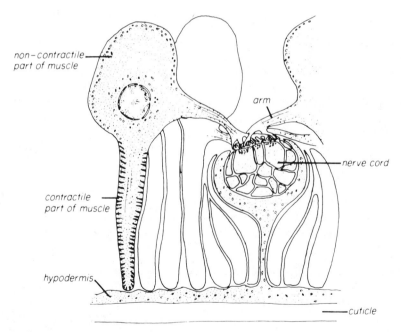

Fig. 2.13 Muscle cells and myoneural junctions in transverse section. The arm of the muscle innervation process subdivides at the junction with the nerve cord. (From D. L. Lee and H. J. Atkinson. 1976. Physiology of Nematodes, 2nd ed. Macmillan, New York, 215 pp., by permission.)

SUMMARY

Nematodes represent an offshoot from the mainstream of animal evolution and have not given rise to any more complex forms of life. Their tubular construction and enclosed body systems seem well suited to life in soil. Both structural and physiological adaptations for different conditions of life are striking. Parasites of insects and vertebrates have evolved different methods to locate, enter, and maintain positions in hosts as well as to permit young of the next generation to repeat the journey. Phytonematodes are specialized to puncture plant cells either from the outside or from within their hosts. Because our subject is phytonematodes and the major environment for these is soil and plants, we must now describe conditions in soil that nematodes encounter.

REFERENCES

1. Bolla, R. 1980. Nematode energy metabolism. In: Nematodes as Biological Models, Vol. 2 (B. M. Zuckerman, ed.). Academic Press, New York, pp. 165 – 192.
2. Croll, N. A. and Sukhdeo, M. V. K. 1981. Heirarchies in nematode behavior. In: Plant Parasitic Nematodes, Vol. 3 (B. M. Zuckerman and R. A. Rohde, eds.). Academic Press, New York, Chapter 9, pp. 227 – 251.
3. Doncaster, C. C. and Seymour, M. K. 1973. Exploration and selection of penetration site by Tylenchida. Nematologica 19: 137 – 145.
4. Dropkin, V. H. and Acedo, J. 1974. An electron microscopic study of glycogen and lipid in female *Meloidogyne incognita* (root-knot nematode). J. Parasitol. 60:1013 – 1021.
5. Dusenbery, D. B. 1980. Behavior of free-living nematodes. In: Nematodes as Biological Models. Vol. I (B. M. Zuckerman, ed.). Academic Press, New York, pp. 127 – 158.
6. Inglis, W. G. 1983. The design of the nematode body wall: the ontogeny of the cuticle. Aust. J. Zool. 31: 705 – 716.
7. John, C. D. and Stretton, A. O. W. 1980. Neural control of locomotion in *Ascaris*: Anatomy, electrophysiology, and biochemistry. In: Nematodes as Biological Models, Vol 1 (B. M. Zuckerman, ed.). Academic Press, New York, pp. 159 – 195.
8. Krusberg, L. R. 1971. Chemical composition of nematodes. In: Plant Parasitic Nematodes, Vol. 2 (B. M. Zuckerman, W. F. Mai, and R. A. Rohde, eds.). Academic Press, New York, pp. 213 – 234.
9. Myers, R. F. and Krusberg, L. R. 1965. Organic substances discharged by plant-parasitic nematodes. Phytopathology 55:429 – 437.
10. Storey, R. M. J. 1984. The relationship between neutral lipid reserve and infectivity for hatched and dormant juveniles of *Globodera* spp. Ann. Appl. Biol. 104:511 – 520.
11. Wallace, H. R. 1973. Nematode Ecology and Plant Disease. Arnold, London, 228 pp.
12. Wyss, U. and Zunke, U. 1986. Observations on the behavior of second stage juveniles of *Heterodera schachtii* inside host roots. Rev. Nématol. 9:153 – 165.
13. Zengel, J. M. and Epstein, H. F. 1980. Muscle development in *Caenorhabditis elegans*: A molecular genetic approach. In: Nematodes as Biological Models, Vol 1. (B. M. Zuckerman, ed.). Academic Press, New York, pp. 73 – 126.

Three

SOIL ENVIRONMENT

Introduction
Structure
Texture
Moisture
Determination of Water Status of Soils
Soil Biota
Summary
References

INTRODUCTION

What are the conditions of life in soil for a nematode? The design of nematodes—their narrow, cylindrical shape, locomotion by undulating movements, the tough, flexible, oily body wall—all the attributes of phytonematode structure and physiology appear to be adaptations to life in soil. These organisms cannot force their way through soil as earthworms do, but must thread through spaces already present, gliding along particle surfaces in films of water. To understand soil nematode environments, we must therefore consider soil microstructure, composition of the soil solution, soil moisture and atmosphere, and soil biota. Until recently soil science did not emphasize conditions in the microworld inhabited by nematodes. Soil science developed from the need to maintain and improve crop production. Engineering requirements for support of heavy structures, for road construction, and for the control of water have also shaped the direction of soil science. Recent availability of new instruments such as the electron microscope and computer-assisted measuring devices for photographs are improving knowledge of soil at the size level of nematodes (1,3,7).

STRUCTURE

Soil is composed of three phases: (a) **solid**—particles resulting from disintegration of rocks at the earth's surface and the remains of organisms living in or on it (matrix); (b) **liquid**—the water; (c) **gas**—the soil atmosphere. Solid portions vary in size, shape, and composition depending on their source and the soil's history. Particles derived from limestone are very different from those of volcanic origin. Rich bottom lands along floodplains of rivers differ from upland areas. Soil that develops under grass is different from agricultural soils. Soils of the tropics are not the same as those of temperate climates. And the cropping history of a field affects its composition and suitability for future use.

> The soil is a heterogeneous, polyphasic, particulate, disperse, and porous system, in which the interfacial area per unit volume can be very large. The disperse nature of the soil and its consequent interfacial activity give rise to such phenomena as adsorption of water and chemicals, ion exchange, adhesion, swelling and shrinking, dispersion and flocculation, and capillarity (4).

The solid phase is called the soil **matrix**; the liquid soil water contains compounds and ions and is called the soil **solution**; the gas phase, soil **atmosphere**, varies in composition depending mostly on water content of the soil.

Above all, soil is a dynamic system undergoing continual change, sometimes rapidly enough to be noticeable in a single season, but more often slowly. Rocks disintegrate by expanding and contracting as the temperature fluctuates, especially when water freezes in cracks. Roots also contribute to formation of fragments. Windborne sand scours surfaces. Many chemical reactions dissolve rock minerals. Water percolates from the surface to the water table, moving minerals and organic matter. Water at the surface may remove soil from one place and deposit it elsewhere, meanwhile sorting components by size and weight. Cycles of freezing and thawing alter soil structure, creating cracks and heaving stones to the surface. Plant roots penetrate all but the heaviest soils and leave voids when they die. Microorganisms cycle organic matter back to simple compounds or fix nitrogen from the air into soluble compounds, and add their own remains to soil. Earthworms and other organisms burrow and move litter to lower horizons, breaking it into smaller particles and digesting parts along the way. Mites and other small animals feed on plants and deposit their feces. And humans, by our agricultural activities, compact soil with heavy machinery and the weight of livestock, then break it up to prepare for crops. We also add pesticides and fertilizer to maintain agricultural production.

Spaces between soil particles are called **pores** or **voids.** They vary in size and shape according to size and composition of the particles. Nematodes live in the labyrinth of interconnecting passages that may be ample or tight. They move in pores of 20 – 30 μm or larger diameters. Pores in a topsoil in good condition can occupy 50 – 60% of its total volume but in a compact subsoil they may occupy only 25 – 30% of the volume. An occasional root forces its way through, in some places pushing particles aside as it grows down and leaves a void when it decays. The nematode, however, cannot make its own passage. Unlike an earthworm, it must reach across the void of a soil pore or move around the side to find an entrance into the next space. Consequently, as pores open and close during the seasonal cycle, nematode populations encounter changing conditions.

TEXTURE

Everyone who handles soil makes a judgement about its texture. Is it a heavy clay or a light sandy soil; does it feel like loam? Will it be a good medium for plant growth? These judgments reflect our perceptions of particle size distributions of soil. A classification can be made to distinguish between soils with the finest particles and those with the coarsest, and to distinguish among various mixtures of particles of different sizes. Particles of 2 μm in diameter ($= 2 \times 10^{-6}$ meters) or less make up **clay. Silt** is composed of particles between 2 and 50 μm, and **sand** grains range from 50 to 2000 ($= 0.05$ to 2 mm). A common system of naming various combinations of particles is based on the various proportions of clay, silt, and sand in a soil. Thus a loam is a soil with 10 – 25% clay, 30 – 50% silt, and 20 – 50% sand. Soils like this, but with more sand are sandy loams. There are also clay loams, silty loams, clay soils, and so on (6). Figure 3.1 presents a diagram of a soil profile with nomenclature of the various layers and some indication of the processes of soil formation.

Soil texture is related not only to the proportions of different sized particles, but also to other constituents. Remains of plants and animals are consumed by bacteria, fungi, small arthropods, and earthworms. The residue, consisting of organic remains and the feces of consumers, is oxidized, and soluble compounds are leached away. The final product (humus), binds soil particles into aggregates (crumbs) that keep a soil porous and favorable for plant growth.

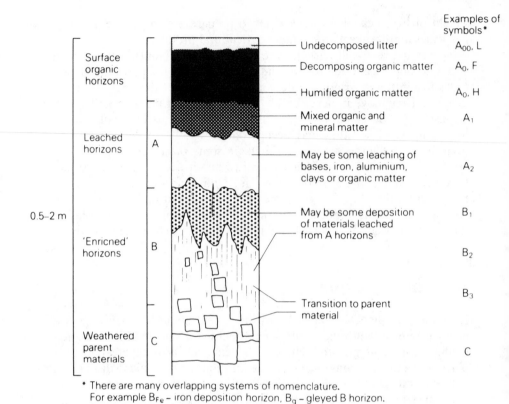

Examples of
symbols*

Undecomposed litter — A_{00}, L

Decomposing organic matter — A_0, F

Humified organic matter — A_0, H

Mixed organic and mineral matter — A_1

May be some leaching of bases, iron, aluminium, clays or organic matter — A_2

May be some deposition of materials leached from A horizons — B_1

— B_2

— B_3

Transition to parent material

C

Surface organic horizons

Leached horizons — A

'Enriched' horizons — B

Weathered parent materials — C

0.5–2 m

* There are many overlapping systems of nomenclature.
For example B_{Fe} – iron deposition horizon, B_g – gleyed B horizon.

Fig. 3.1 Diagram of the soil profile and its horizons. (From K. Simpson. 1983. Soil. Longman Handbook in Agriculture, New York, 238 pp., with permission.)

MOISTURE

Clay is the most important constituent of soil in relation to microstructure. By its surface properties, clay affects behavior of water and of solutes. The amount of surface present in soil is astonishing. Thus 1 g (dry weight) of clay has from 5 to 800 m² of surface, depending on the type of clay. Silt and sand contribute much less surface to the total. Whereas sand and silt are made of minerals originally present in parent rock, clay incorporates secondary minerals derived from the primary ones. Clay minerals in cooler soils are crystals of aluminum combined with silicon and water (hydrated aluminosilicates). Clay in tropical soils consists of oxides of iron and aluminum combined with water molecules. Clays form platelike structures that stack together in layers and bear negative surface charges. Both minerals in solution and water itself bind to surfaces of the stacks. The surface also influences aggregation of dispersed particles into clumps of various sizes, and the electrical charges may influence nematode behavior. Clay swells when wet and shrinks as it loses water, thus causing soil to crack during drought.

After rainfall has filled all pores in a soil, water begins to move downward by gravity. Pores wider than 3 mm drain away rapidly, followed by slower drainage from smaller pores (30 μm – 3 mm). Water can move in the soil in all directions. Water in the small pores (< 30 μm) is held by capillary action. Eventually drainage by gravity stops and the soil is said to be at **field capacity.** Now gravity is balanced by forces retaining water. As a soil continues to lose water to the atmosphere by evaporation and as plants take up water through roots, the forces opposing water loss increase. Nematodes are affected by these forces because as soil dries, water retreats from progressivly smaller pores and nematodes are denied access to dry pores. Furthermore, when films of water become very thin, surface tension gets strong enough to trap nematodes in small droplets. Figure 3.2 indicates the relation between soil texture and the waterholding capacity of soils.

Because texture affects pore size and thus water retention, knowledge of soil water content by itself gives little information about soil conditions. Knowledge of **water potential** is more useful. Water potential is the amount of energy holding water in soil and consequently the energy needed to remove water from soil. This can be described in several ways. Amount of pressure to extract water from soil is a measure of water potential. The **bar** is one unit of measurement. One bar denotes pressure approximately equal to that of 1 atm at sea level at the temperature of melting ice. Water potential is often stated in agriculture as bars of tension (also called suction), that is, the force needed to remove water from a soil in a given condition of dryness. Plants usually cannot remove water from soils at values greater than 15 bars of tension. Water in the finest pores is unavailable to plants. Another usage is to state the force holding water as **negative bars.**

In addition to bars, another way to express force needed to overcome tension holding water in soil is to state the **height of water** in a column that exerts this amount of force. Thus in a soil at −1 bar it takes 1 atm of pressure to extract water. This is equivalent to the pressure of a column of 1000 cm of water (10 m). To avoid the large numbers for dry soils, one can use the logarithm of the centimeters of water. This number is called the **pF** of a soil and many researchers use this system to characterize soil tensions. Thus soil at a pF of 3 is at −1 bar of suction.

Fig. 3.2**a** is a plot of the tension in two soils in relation to water content. At 15 bars tension, the clay loam has 14% moisture, but the sandy loam

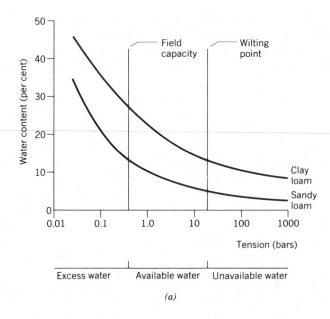

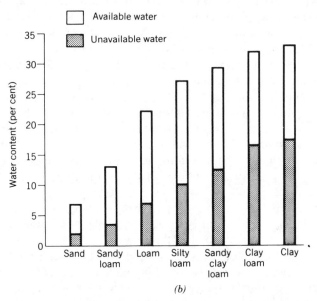

Fig. 3.2 Texture – water relations. (a) The relationship between soil texture, water content, and water tension; (b) soil texture, available water capacity and unavailable water capacity. (From K. Simpson. 1983. Soil. Longman Handbook in Agriculture, New York, 238 pp., with permission.)

has only 5%. In other words, it takes more energy to remove water from clay soils than from sandy soils. This is also illustrated in Fig. 3.2**b**. Here distinction is made between available and unavailable water. Unavailable water is held in small pores or is bound to soil particles, especially clay.

DETERMINATION OF WATER STATUS OF SOILS

The **moisture characteristic** of a soil is a graph of suction (negative bars) against moisture content. The latter is determined by loss of weight when soil is held at 105°C for 24 hr. This temperature is sufficient to drive off all the water in a soil. Suction is measured by tensiometers or in other ways. The **tensiometer** consists of a porous plate in contact with soil on one side and with water on the other. Figure 3.3**a** shows a simple apparatus by which known suctions may be applied to samples of soil. This kind of tensiometer has been used to study changes in nematode behavior under increasing suctions. Another tensiometer is illustrated in Fig. 3.3**b**. This instrument is used in the field to determine when to apply irrigation. It consists of a ceramic cup joined to a sealed tube containing water. A vacuum gauge connected to the tube indicates suction. The instrument is placed in a snug hole and the soil is firmly tamped so that the cup presses tightly against the soil. Water in the sealed column reaches equilibrium with that of the soil and the amount of suction is shown on the dial.

The **pressure membrane** apparatus consists of a chamber with a membrane upon which a soil sample is placed. The chamber is then sealed, gas pressure is applied, and the amount of water remaining in the sample at various pressures is determined. The **thermocouple psychrometer** is an instrument that measures relative humidity of the soil atmosphere. A thermocouple placed in soil is cooled by passing an electric current through it for about 30 sec. Water from the soil atmosphere condenses on the cooled surface in the same way as dew is deposited on leaves during cool nights. The condensed water then evaporates from the thermocouple and cools it, creating a small, temporary voltage. The amount of voltage indicates relative humidity of the soil atmosphere, from which soil suction is calculated.

It is often necessary to adjust soils to specific moisture content in experimental work and the **moisture characteristic** is useful here because from it one can calculate the size of pores just draining at a particular tension. The diameter of pores just emptied as a soil dries is calculated by the formula

$$D = 3000/h \ \mu m$$

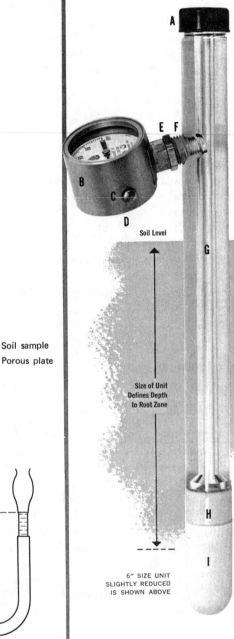

Soil sample

Porous plate

h

(a)

A

E F

B

C

D

Soil Level

G

Size of Unit
Defines Depth
to Root Zone

H

I

6" SIZE UNIT
SLIGHTLY REDUCED
IS SHOWN ABOVE

(b)

(A) Plastic filler cap protects internal neoprene "O" ring vacuum seal from wind blown soil particles and shields from sun and rain.

(B) Hermetically sealed Bourdon Dial Gauge is completely weather proof. The exceptional sensitivity of these gauges results in unusually short response time. A neoprene "O" ring stem seal permits the dial gauge to be removed or oriented in any convenient reading position, with perfect vacuum seal assured; and thus provides desirable flexibility for transportation and use of the instrument.

(C) Vent screw, used to equalize the pressure inside the case with atmospheric pressure at your elevation.

(D) Neoprene diaphragm on bottom side of gauge case provides compensation for daily temperature-pressure changes of air within the gauge case.

(E) Lock nut secures the dial gauge in any desired orientation.

(F) Weather-proof plastic connection fitting, integral with body tube is positively leak free.

(G) Heavy wall, clear plastic body tube, 7/8" O.D. is immune to weather effects and soil chemical action.

(H) Porous ceramic cup is bonded directly to body tube without the use of cement to make a vacuum-tight, lifetime joint.

(I) Carefully manufactured porous ceramic cup provides maximum flow rate through cup wall for sensitive response to changes in soil moisture.

Cat. No. 2002 Insertion Tool can be used to core hole in soil to provide clearance for body tube and proper hole diameter for ideal contact between porous cup and soil.

The easy reading dial, shown actual size, is graduated 0-100 centibar of soil suction, which is the standard unit of measurement.

where D = pore diameter; h = height of a column of water whose pressure equals that of the soil suction. Thus a column of water 1000 cm high exerts a pressure approximately equal to atmospheric pressure. Water will be removed from a soil by this amount of suction until the interfaces reach pores with a diameter of 3 μm.

Figure 3.4 illustrates several aspects of the soil moisture characteristic. The diagram at the top shows the appearance of pores at different tensions. The vertical axis has scales for moisture content and also percentage of pore space containing water. The horizontal axis shows suction and also diameters of pore necks just drained of water.

Wallace (8) demonstrated that nematode activity is related to the position on the moisture characteristic curve that a soil occupies when nematodes are present. The most favorable moisture for activity of some nematodes was at the point where the curve changed from convex to concave (5). The particular moisture characteristic curve for any soil is related to its texture. A clay soil has a moisture characteristic curve that is not as steep as that of a soil of coarser texture. Moreover, the curve obtained while a soil is drying differs from that of a soil undergoing water uptake.

The discussion until now has treated **matric potential** of water in soil. We have considered energy relations resulting from capillary forces that reflect the effect of water surface tension and sizes of the pore openings (pore necks). We have indicated also that adsorption to clay particles restricts water movement. A third component of the total picture is **osmotic pressure** of the soil solution. This is usually less important than matric potential and adsorption. In saline soils, however, compounds in solution also diminish free energy of water to evaporate, be taken up by plant roots or move in the soil.

In addition to water potential, two other parameters are important to nematodes as well as to roots. Nematodes cannot control their body **temperature**. Thus their activity depends on soil temperatures. Below about 10°C, nematodes are not very active and usually pose no threat to plants. A few crops can grow at temperatures below this. For examples, some early crops of potato are planted before the soil has warmed above 10°C, and by the time nematodes become active, the plants have developed substantial roots. When the soil warms and the nematodes invade roots,

Fig. 3.3 Tensiometers. (a) Simple tensiometer; *h* is the height of water column applying suction to soil (From D. M. Griffin. 1972. Ecology of Soil Fungi. Syracuse Univ. Press, Syracuse, New York, by permission of Chapman & Hall). (b) A commercially available tensiometer (irrometer) used in management of irrigation.

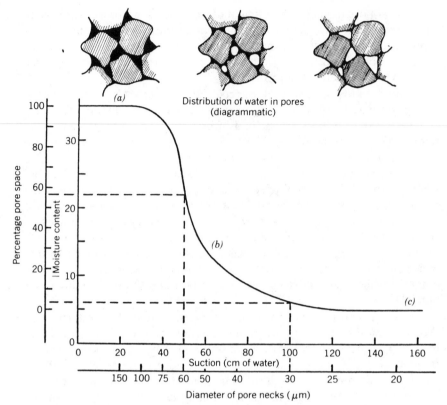

Distribution of water in pores
(diagrammatic)

Fig. 3.4 Diagrammatic illustration of the moisture characteristics of a hypothetical soil with pores (**a**) water-filled, (**b**) emptying, and (**c**) empty — shown in relation to moisture content and suction pressure. (From F. G. W. Jones. 1959. Ecological Relationships of Nematodes. In: Plant Pathology, Problems and Progress, 1908 – 1958, Univ. of Wisconsin Press, Madison, published for the Amer. Phytopathol. Soc., C. S. Holton et al., eds., pp. 395 – 411, with permission of the Regents of the Univ. of Wisconsin.)

the plant has already made sufficient growth to tolerate the damage. Most plants, however, are adapted to soil temperatures in the range at which nematodes are active.

Oxygen content of the soil atmosphere also affects nematodes and roots. Oxygen diffuses rapidly from the atmosphere above ground into the soil atmosphere but more slowly from this into soil water where the nematodes are. Water film thickness controls the rate at which oxygen is available to nematodes and roots. An instrument (the platinum microelectrode) measures diffusion rates into films of water. It provides a more realistic measure of soil conditions than can be obtained by measurement of soil atmosphere. With this instrument nematode response to low rates of oxygen diffusion has been determined. When the rate falls below a critical level, the nematodes die or decrease their reproduction. This level is close to that at which root growth is inhibited.

We should now realize that a complex system of interactions operates in soils. Texture determines the distribution of pore spaces; moisture changes conditions within pores. Organic matter influences particle aggregation into crumbs, thus enlarging the effective pore size. Growing and decaying roots affect soil properties. Most roots exude a mucilaginous gel that binds soil particles and provides an excellent substrate for microorganisms. The microorganisms consume oxygen and release CO_2 and other products. Nematodes have physiological limits that are sometimes exceeded in this environment. Their defense is usually to enter a state of physiological inactivity until conditions change for the better or they die. Nematode successes, although they may cause us great problems, are indeed remarkable when viewed against the environment in which they live.

SOIL BIOTA

As mentioned already, soil is a dynamic system undergoing change at all times. Many kinds of organisms are the agents of change. Soil microorganisms include bacteria, algae, fungi, actinomycetes, protozoa, and rotifers. Somewhat larger are the nematodes, mites, and small arthropods. Annelids, molluscs, and arthropods are larger still, and burrowing mammals that spend their entire lives underground or those that simply nest underground top the list. In addition to all of these, roots of crops and of weeds plus roots of shrubs and trees are part of the life in soil.

You can measure the total metabolism of these organisms in a core of soil by following CO_2 production in a suitable apparatus. What are they all doing in soil? In common with all forms of life they are feeding, reproducing, excreting, responding to environmental cues, and dying. Organisms affect soil in various ways:

1. They degrade organic matter that falls on the surface and is brought to the interior by their activities.
2. The mucilaginous materials secreted by many algae, roots, and earthworms bind small particles into aggregates, thus increasing soil porosity.

3. Excretions from living cells contribute nutrients to the soil solution and may acidify it. In some cases acid soil solutions attack the parent rocks.

4. Earthworms alter soil components by ingesting them; the feces of mites and other small arthropods are deposited in quantity in many soils. These products become part of the highly favorable humus that improves soil texture.

5. Respiration of plants and animals depletes oxygen from soil atmosphere. In saturated soils anaerobic conditions prevail soon after connection to the atmosphere above ground is interrupted. In well-drained soils there are also localized regions, such as spaces within soil crumbs, with low oxygen.

6. Appreciable quantities of nitrogen and other elements are sequestered in microorganisms as well as in larger species. Fertilizers applied to crops are diverted in part to the soil biota. Of course this pool of nutrients turns over when the soil flora and fauna die.

7. Bacteria and algae fix atmospheric nitrogen, thus contributing to soil fertility.

Algae are described briefly here because they are important in soil, but few nematologists are aware of them. Algae are among the first organisms to colonize soil during its formation. They create conditions that permit other forms of life to succeed them. Algae are single cells or colonies of cells containing chlorophyll and without separate tissues forming roots, stems, and leaves. Marine forms such as the familiar seaweeds may be large, but species in soil are mainly single-celled or develop small filaments. Chlorophyll functions in light; therefore algae thrive on the soil surface. But they also occur in deeper layers down to 1 – 2 m. Algae are most numerous at about 7 – 15 cm depth, with diminishing numbers below this.

Algae of cultivated soils include many species representing several taxonomic groups. Some species have been grown in the laboratory in the dark with nutrient media that supply sources of energy, but it is doubtful that much growth takes place within soil. They are mixed with soil during cultivation, are carried down by animals, seep down with water, or move under their own power. Many algae produce motile spores that can swim just as protozoa do.

This group of organisms faces frequent changes of soil moisture. They have outstanding resistance to desiccation. Some live algae were recovered from soils that had been kept dry in England for 79 years! Under rapid desiccation, many individuals die but some survive, and the species continues. Under extreme and prolonged drought, resting cells of algae within soil survive. They also survive extreme cold.

A few species fix nitrogen from air, but quantitative data on the importance of algae in the nitrogen cycle of soil are scarce. All algae release oxygen as a result of photosynthesis. They are important in rice cultivation under water because they aerate the water and supply nitrogen to the crop.

Many species of nematodes apparently feed on algae. The evidence is not extensive. We know that some types of nematodes with a particular kind of spear have been cultured on algae growing on agar (2).

Fig. 3.5 SEM of undisturbed clay, air-dried, vertical fracture (×108750). (From P. Smart and N.K. Tovey. 1981. Electron Microscopy of soils and sediments: Examples. Oxford University Press, New York, 178 pp., 264 pp., by permission.)

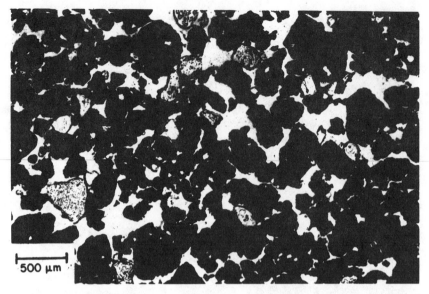

Fig. 3.6 Thin section of soil showing granular structure composed of subspherical aggregates with frequent pore spaces (×28). (From E. A. FitzPatrick. 1984. Micromorphology of Soils. Chapman & Hall, New York, with permission.)

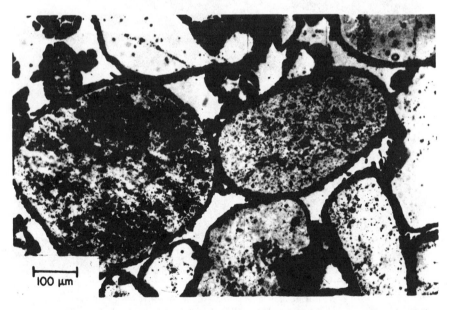

Fig. 3.7 Sand grains bound by bridge structure formed by humus coating (×120). (From E. A. FitzPatrick. 1984. Micromorphology of Soils. Chapman & Hall, New York, with permission.)

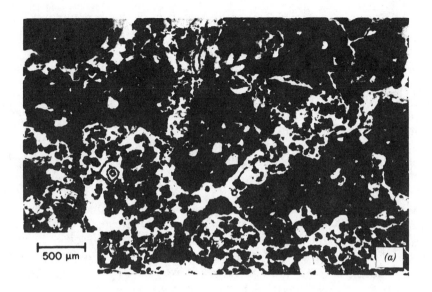

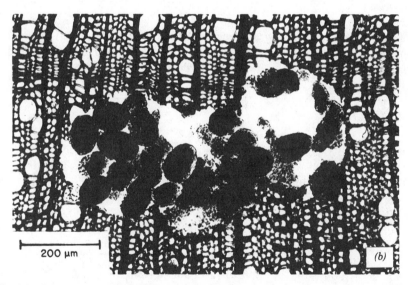

Fig. 3.8 Fecal material in soil. (a) Small beaded and granular fecal material probably of enchytraeid worms replacing earthworm tunnels. An example of the interplay of the soil fauna (×28); (b) Fecal pellets of mites replacing the softer central tissue of plant material. This is a common feature in the organic horizons of many soils (×120). (From E. A. FitzPatrick. 1984. Micromorphology of Soils. Chapman & Hall, New York, with permission.)

Populations of soil organisms are related to each other in complex ways. Life cycles are short so that great numbers of bacteria quickly develop around the dead body of an earthworm, for example. Many species subsist by predation or by parasitizing other forms. Nematodes feed on bacteria, fungi, algae, roots, and each other. In turn, they are victims of mites and small arthropods such as Collembola and of bacterial and fungal parasites and predators. The application of a biocide to soil destroys the equilibrium among all species so that unexpected population shifts often occur. For example, some relatively unimportant phytonematodes in a particular soil reproduce freely after soil fumigation and reach damaging population levels.

Knowledge of soil organisms is uneven. Although specialists on various groups have studied the flora and fauna found in soil, many groups are poorly known. Moreover, it is extremely difficult to investigate the complex relations among soil organisms and especially the relation of soil organisms to crop production. In the future this subject will receive more attention as agriculture becomes more intensive all over the world.

Figures are presented to illustrate various aspects of micromorphology of soils. Figure 3.5 is a highly magnified scanning electron micrograph of air-dried clay. It shows the platelike clay particles. The bar below the figure represents 1 μm. Figure 3.6 illustrates a highly porous, well-drained soil. Figure 3.7 shows how aggregates are formed by binding action of organic matter. In Fig. 3.8, the important role of fecal pellets in soil formation is presented. This is a common feature in organic horizons of many soils.

SUMMARY

Imagine that you are a nematode in soil, seeking a plant root for lunch. You can't see anything, you have no hands or feet. You can detect chemical and temperature stimuli, possibly vibrations. You can move forwards and backwards like a snake, you can stop and turn your head to change direction or to test stimuli from various directions. You may have to go a long way to find the root, and you will know when you are getting closer, and speed up your motion. Eventually you may find a root before you run out of stored energy. Much of the time you will lie inactive, or curl into a tight spiral to avoid losing your body water. You may also have some frightening encounters. If you put your head into a loop of fungus, you may be lost, the loop will spring shut, and filaments will invade your body to consume your insides, no matter how hard you thrash about. Or a spore of a bacterium may land on your shiny, slick cuticle and stick to it like a plastic bomb sticks to the side of a submarine. From this spore, a filament will enter your body and eventually destroy you. Or a roving mite may chance to meet you and punch a hole into your plump side, then suck out the contents. If you fail, or are killed, don't worry. Some of your close relatives will succeed and keep the species going.

REFERENCES

1. Bisdom, E. B. A. and Ducloux, J. 1983. Submicroscopic studies of soils. Elsevier, Amsterdam. Dev. Soil Sci. 12: 356 pp.
2. Chapman, V. J. 1968. The Algae. Macmillan, New York.
3. FitzPatrick, E. A. 1984. Micromorphology of Soils. Chapman & Hall, New York, 433 pp.
4. Hillel, D. 1980. Fundamentals of Soil Physics. Academic Press, New York, 413 pp.
5. Jones, F. G. W. 1959. Ecological Relationships of Nematodes. In: Plant Pathology, Problems and Progress, 1908 – 1958 (Holton, C. S. et al., eds.). Univ. of Wisconsin Press, Madison, published for the Amer. Phytopathol. Soc., pp. 395 – 411.
6. Simpson, K. 1983. Soil. Longman Handbook in Agriculture, London, 238 pp.
7. Smart, P. and Tovey, N. K. 1981. Electron Microscopy of Soils and Sediments: Examples. Oxford Univ. Press, Oxford, 178 pp.
8. Wallace, H. R. 1958. Movement of eelworms. I. The influence of pore size and moisture content of the soil on the migration of larvae of the beet eelworm, *Heterodera schachtii* Schmidt. Ann. Appl. Biol. 46:86 – 94.

Four

METHODS

INTRODUCTION

Faced with a diagnostic problem, an investigator begins by isolating possible culprits from diseased plants and soil. The phytonematologist therefore extracts nematodes to estimate numbers and to prepare them for identification under the microscope. This chapter describes the principal available methods.

Phytonematodes are obligate parasites with relatively slow life cycles; therefore culture techniques of microbiology cannot be used. Nematodes within plants can be stained in place or separated from tissues. Nematodes in soil must be separated from the sand, silt, and clay particles and from organic debris. Soil must be dispersed in water to free nematodes, which can then be caught on sieves. Alternatively, the organisms can be permitted to crawl out of saturated pores and fall into water from which they can be collected. After separation from plants and soil, nematodes must be prepared for examination with the light microscope. Living specimens can be anesthetized or fixed in solutions to preserve their structures with little distortion. Because nematodes have a high proportion of water in their tissues special methods are required for permanent mounts. They tend to shrink and distort unless water is gradually replaced with glycerol. In recent years special techniques for use of the scanning electron microscope have been developed.

STAINING NEMATODES IN PLANT TISSUES

Acid Fuchsin

1. Prepare a solution of acid fuchsin by adding 25 mL glacial acetic acid to 75 mL deionized or distilled water. Dissolve 0.35 g acid fuchsin in the solution.
2. Wash roots or other plant tissues to remove adhering soil particles and cut them into short pieces (1 – 2 cm).
3. Bleach roots in dilute sodium hypochlorite solution for about 4 min with occasional agitation. The exact strength of bleach and time required must be determined by trial and error. Young roots may be tested with 10 mL bleach (5.25% NaOCl) + 50 mL water; older roots may require 20 mL bleach + 50 mL water, and so on.
4. Rinse the roots well in running tap water and soak in water for at least 15 min to remove bleach. Then drain and place roots in a heat resistant container with 50 mL water + 1 mL stain solution. Boil solution for 30 sec in a ventilated area to protect yourself against vapors of acetic acid.
5. Cool, drain, and rinse roots in running tap water.
6. Transfer roots to acidified glycerol (30 mL glycerol + 2 or 3 drops of 5 n HCl). Heat to boiling and cool. Roots will be clear, nematodes red.
7. Store in acidified glycerol. For examination mount tissues on microscope slides, cover, and subject to gentle pressure. Or place them in a small amount of acidified glycerol in a Petri dish cover. Then apply gentle pressure with a Petri dish bottom, and count the nematodes under a dissecting microscope ($\times$ 40 magnification). This method of staining is also applicable to aerial parts of plants [(adapted from ref. 10)].

For treatment of lignified tissues, use stronger NaOCl solutions or increased time of treatment, or soak in 30% H_2O_2 for 1 hr, then rinse and stain as above (2).

Phloxine B for Staining Eggs in Egg Sacs

Roots bearing egg sacs of *Meloidogyne* spp. may be soaked in a solution of phloxine B (150 mg/L) for 15 min or longer. Eggs readily take up the purple stain (13).

EXTRACTION FROM PLANT TISSUES

A. *Soaking in Water*. Tissues should be free of soil and cut into short segments (5 – 10 cm lengths). Place plant materials into containers that can be tightly closed. Add water sufficient to cover the plant segments. Seal the container and incubate at room temperature. Migratory **endoparasites** will leave the roots and can be collected on a sieve with openings of 38 μm or smaller. Add water to the container each day, shake to wash the roots, and decant through a coarse sieve (250 μm openings) over the fine one. Add fresh water and collect nematodes daily until no more emerge. Addition of hydrogen peroxide to the water will improve extraction by raising the level of dissolved oxygen and antibiotics will inhibit growth of microorganisms.

B. *Mechanical Maceration*. Pieces of plant tissue 2 – 3 cm long macerated in water in an electric blender for 15 – 30 seconds will produce a mixture of living nematodes and fragments of plant tissue. This mixture can then be distributed on a sieve (38 μm openings). The sieve is maintained in water just sufficient to cover the debris. Motile nematodes will make their way through openings in the sieve and may be collected from the water below. Nonmotile stages of endoparasites often fragment in this procedure but some intact specimens can be found in diluted portions of the mixture.

C. *Enzyme Maceration*. Enzymes that macerate plant tissues, such as pectinases, cellulases, and hemicellulases, can be obtained commercially or from cultures of soft-rotting bacteria and fungi. Details of optimal conditions for use are specific for the particular preparations (5).

D. *Agitation in a Shaker*. Nematodes will also emerge from plant tissue in water under continuous agitation in a motor-driven shaker. A wrist-action shaker provides sufficient agitation to ensure adequate aeration and some maceration. The system can be operated continuously for days.

E. *Intermittent Mist*. Plant materials are supported on coarse plastic screening in a funnel that drains into a large test tube or pan. A spacer between the funnel and test tube is required to prevent flooding of the funnel. Water under pressure flowing through small orifices creates mist. A solenoid in the line connected to a time clock provides intermittent mist. As nematodes reach plant surfaces, water from the mist washes them down the funnel. The supply of water is regulated to move nematodes without washing them out of the receptacle. This system can be used for several days to collect large numbers of nematodes from appropriate plant materials.

EXTRACTION FROM SOIL

Nematodes are never distributed uniformly in soil. They are usually more numerous close to plants, and they may be less numerous in certain parts of a field because of wetness, soil type, or other factors. It is therefore important to sample a field of interest carefully to obtain a representative portion of soil. One usually takes soil cores from a number of places, bulks them together, and mixes thoroughly. Then subsamples are removed for examination. A tool commonly used for obtaining soil samples is shown in Fig. 4.1. Sampling strategies are discussed in Chapter 8.

Nematodes may be separated from soil in several ways. Each method should be thoroughly tested, preferably with respect to its efficiency, before adoption as the standard for a laboratory.

A. *Migration out of Wet Soil: Baermann Funnel.* This method, although inefficient, is convenient and can be used to detect the presence of active phytonematode species or juvenile stages of sedentary parasites. Wrap about 100 cm^3 of soil in paper tissue or cloth and support it on a coarse plastic screen in a funnel fitted with a rubber tube and a clamp. Pour water gently along the side of the funnel until it just covers the lower surface of the soil. After about 24 hr, open the clamp briefly to collect a small quantity of fluid from the funnel into a watch glass or small beaker. It will contain nematodes that have moved out of the soil and fallen to the bottom of the funnel. Active nematodes are recovered but inactive species are not. Figure 4.2 illustrates a Baermann funnel and a modification for increased efficiency.

Many modifications to improve the original design have been published. Their purpose is to increase oxygenation of the water or to stimulate nematodes to greater activity. One convenient system is to distribute soil in a shallow layer 1 cm or less deep over paper or cloth supported by a coarse sieve. The sieve is set into a shallow pan in which the water level just reaches the soil. Nematodes recovered after migrating out of soil may be

Fig. 4.1 A soil sampler for removing cores of soil. (Photo courtesy of Soiltest, Inc.)

concentrated on a fritted glass filter, by sedimentation in conical vessels, or by centrifugation. Avoid the use of copper sieves, which may be toxic to the nematodes.

B. *Cobb's Sieving and Gravity Method*. The specific gravity of nematodes (about 1.05) is just above that of water, so that they settle from suspension more slowly than soil particles. In this method, soil is suspended in water and permitted to settle for a brief period, and the water containing nematodes is passed through sieves. The following procedure, adapted from Ref. (7), may be followed:

1. Place a soil sample of 100 cm³ in a bucket and add 2 or 3 L tap water.
2. Stir with a stick until all clods are broken up. Let the mixture settle without disturbance for 30 sec to 1 min.
3. Pour the mixture into a second bucket through a coarse sieve (openings of 0.35 – 0.85 mm), leaving heavy soil particles in the first bucket.
4. Repeat steps (2) and (3) with about 1 L water. Most of the nematodes in the soil sample will now be in the second bucket.

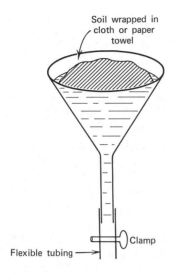

Soil wrapped in
cloth or paper
towel

Clamp

Flexible tubing

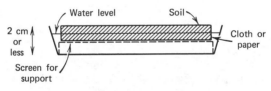

Water level   Soil

2 cm
or
less

Cloth or
paper

Screen for
support

Fig. 4.2 Diagram of standard Baermann funnel and a modification for increased efficiency.

5. Wash the residue on the coarse sieve into the second bucket with additional water to recover nematodes still present on the sieve.

6. Discard the residue on the coarse sieve and in the first bucket and clean these thoroughly.

7. Pour the water from the second bucket gently through a fine sieve (openings of 0.063 mm or smaller) held at an angle to reduce the effective size of the openings. Avoid pouring the mud at the bottom of the bucket through the sieve. Wash the fine sieve with a gentle stream of water to remove fine particles.

8. Invert the fine sieve at an angle greater than vertical over a container, (for example, a 100 cm^3 beaker or a drinking glass) and wash nematodes off the screen with about 25 cm^3 of water poured from behind the screen.

9. Steps (7) and (8) may be repeated by catching the water in the first bucket.

10. The nematodes may be further concentrated by permitting the back-wash from the fine sieve to settle for 30 min, then carefully decanting all but the last centimeter of water. An experienced operator can recover a large percentage of the nematodes in a soil sample in about 10 min. But the procedure must be carefully done. If the nematodes at the end of the process are mixed with unacceptable amounts of debris, they may be recovered by sugar flotation (see procedure C) or by permitting them to migrate through fine cloth or paper. A method to free nematodes from debris employs an inverted beaker over a funnel as shown in Fig. 4.3.

C. Centrifugal Flotation. The principle is to separate nematodes from soil in water, followed by centrifugation in water, and finally by centrifugation in a solution of sufficient specific gravity to cause the nematodes to float to the top. The directions are slightly modified from Ref. (12).

1. Mix 100 cm^3 of soil with water to reach a volume of 800 mL.
2. Stir vigorously and allow the soil to settle for 60 sec.
3. Decant over a moderately coarse sieve (openings 0.425 mm) over a fine sieve (openings of 0.038 mm). Rinse the debris on the nested sieves.
4. Wash the debris and nematodes from the fine sieve into a 150 mL beaker, agitate the mixture, and pour the fluid with its contents into 50 mL round bottom centrifuge tubes.

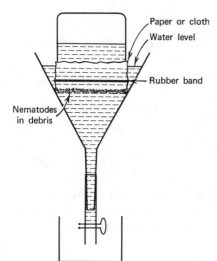

Fig 4.3 Inverted beaker in funnel to permit nematodes to migrate free of debris.

5. Balance tubes in a horizontal rotor and centrifuge at 400 × g for 5 min. Permit the rotor to come to rest without braking.

6. Decant fluid without disturbing the pellet with nematodes at the bottom.

7. Add enough sucrose solution to the tubes to reach halfway to the top. Resuspend the pellet with a glass rod and add more sucrose solution to reach 0.5 cm from the top. Rinse the rod after each use. The sugar solution is made by mixing 454 g of sucrose with enough water to make 1 L.

8. Centrifuge at 400 × g for 60 sec and let it come to rest without braking.

9. Decant the sucrose solution over a very fine sieve (openings of 0.028 mm) without disturbing debris at the bottom of the tubes. With 30 mL water gently rinse the contents of the sieve into a beaker.

This method recovers most nematodes from soil, including both active and inactive forms. For best results, the soil should contain some clay to hold the nematodes in the pellet in step (6). If clay is lacking, add a small amount of kaolin to the fluid at step (4).

D. *Elutriation*.   Several different devices are in use that precisely regulate the rate of water flow to bring nematodes up in the apparatus while soil falls to the bottom. The design of Oostenbrink, described in Ref. (14), is illustrated in Fig. 4.4.

Construction is of rustproof metal. A sample of 100 cm^3 of moist soil is placed on a sieve with openings of 1 mm. Soil is washed into the conical part by a current of water approximately 700 cm^3/min, delivered through a spray nozzle. This current is terminated when all soil has passed through the screen. At the beginning of the operation, the water level in the lower part of the apparatus is adjusted to submerge the lower end of the funnel. At the same time a current of water of 800 cm^3/min enters the base through the tube while soil is being washed into the apparatus. The current is subsequently reduced to 400 cm^3/min. A flow meter in the line is useful. This current prevents nematodes and fine soil particles from entering the lower, narrow portion of the elutriator while the heavy particles sediment to the bottom. The operation, which lasts 10 – 15 min, is halted when the water level reaches the line shown at (14). The plug is removed and the entire contents of the upper part of the elutriator drain through the nest of four sieves with openings of 0.045 – 0.050 mm.

Different special purpose elutriators for collection of cyst nematodes from soil have been built.

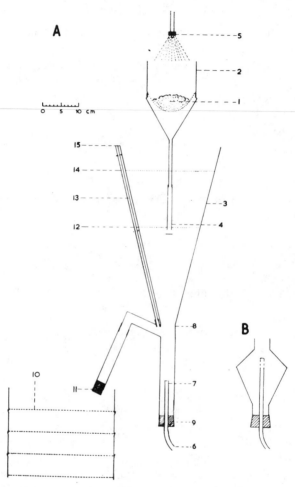

Fig. 4.4 Flotation apparatus for extraction of active nematodes from soil. (**A**) for normal, (**B**) for large samples. (1) Sample of moist soil; (2) 1 mm top sieve; (3) Flotation apparatus; (4) Funnel-pipe with baffle plate; (5) Nozzle of about 700 mL of water per minute; (6) Upward water current; (7) Insertion pipe with perforated tip; (8) Neck of the apparatus where the nematodes accumulate; (9) Rubber plug to be removed for flushing the apparatus; (10) Nest of sieves; (11) Rubber plug of the outlet; (12) Water level at start; (13) Plastic gauge; (14) Water level when nozzle 5 is closed and the upward current is adjusted to 400 mL/min; (15) Water level when plug 11 is pulled out. (From M. Oostenbrink. 1960. Estimating nematode populations by some selected methods. In: Nematology, Fundamentals and Recent Advances with Emphasis on Plant Parasitic and Soil Forms (J. N. Sasser and W. R. Jenkins, eds.). University of N. Carolina Press, Chapel Hill, pp. 85 – 102, by permission.)

Sieves of graded sizes may be purchased from scientific supply houses. Stainless steel construction is preferred. Inexpensive substitutes may be fabricated from sections of plastic pipe and cloth fitted together like embroidery hoops. Nylon mesh of accurately known mesh size is available. Plastic cups cut transversely in half also nest together nicely for constructing sieves.

PREPARATION OF NEMATODES FOR MICROSCOPY

Because nematodes from soil or plants are usually too small to be handled without magnification, preparation of specimens is begun under a dissecting microscope with the organisms in water in a small glass (Syracuse) or plastic dish. Suitable microtools for picking up individual specimens may be made from a bristle of a toothbrush cemented to the end of a glass or wooden rod. "Excellent picks can be made by cutting a section of a dry bamboo cane about 15 cm long and 2 or 3 cm in width. One end of this is tapered to a fine point by rough trimming with a knife. The point is then placed under the low power of a dissecting microscope for further trimming with a razor blade, ending with a point which consists of only a few fibers, preferably outside fibers of the bamboo cane" (17). Micropipettes controlled by mouth suction through rubber tubing are useful. They can be drawn in a gas flame from commercial soft glass pipettes.

A. Living Specimens.   Details of morphology can best be seen on living specimens. The microscope should be of good to excellent quality and the observer must be skilled in its use. An oil immersion lens is often required. A superior source of light is needed, preferably one that can be used in Köhler illumination. The beginner should prepare pencil sketches of observations. Many of the characters required for accurate identification are hard to see. At first one cannot believe that other nematologists have actually seen what they have recorded in their drawings!

Several anesthetics will immobilize nematodes without killing them. Propylene phenoxital in 0.5 % aqueous solution made up fresh every few days is excellent. Another anesthetic is dichlorodiethyl ether (2 drops of ether to 50 mL water in a stoppered bottle) (9). Use of chloretone (2 g in 500 mL water), and of chloroform vapor are described in Ref. (3).

Nematodes may be picked up individually and transferred to a drop of water or anesthetic on a slide. If you wish to avoid pressing on the specimens, add two or three short lengths of glass fibers of appropriate thickness. Cover with a cover glass, remove excess fluid from the preparation, and seal with fingernail polish or other quick-drying sealant around the rim. Such preparations will remain in good condition for several hours.

B. Permanent Mounts—Fixatives. Specimens must first be fixed to precipitate proteins for preservation of specimens in lifelike appearance. Fixatives are permitted to act for 24 hr or longer. Specimens may be kept indefinitely in fixatives, but they may undergo unfavorable changes after prolonged storage. Solutions commonly employed are:

Formaldehyde: 2 – 4% in water.

Formal – acetic: 40% formaldehyde (formalin)   10 mL
       glacial acetic acid   1 mL
       distilled water—make up to   100 mL

Best used at 100°C. Bring nematodes in water to a volume of about 1 mL, heat fixative in a tube held in boiling water, add approximately 5 mL hot fixative to the nematodes.

Formal – propionic: same as formal – acetic except that propionic acid is substituted for acetic acid.

TAF: 40% formaldehyde (formalin)   7 mL
      triethanolamine   2 mL
      distilled water   91 mL

Formal-glycerol: 40% formaldehyde -(formalin)   10 mL
      glycerol   6 mL
      distilled water   84 mL

Heat fixative to boiling; bring nematodes to a small volume of water. Add approximately an equal volume of boiling fixative to the volume of water with nematodes. This causes nematodes to straighten. Consult Ref. (9) for a general discussion of fixation.

C. Permanent Mounts—Transfer to Glycerol. For permanent slides, water must be replaced with a solvent of appropriate refractive index to reveal details of structure. But this replacement must be made without causing specimens to shrink. A slow method, taking weeks, permits evaporation of water from a mixture of water and glycerol, thus infiltrating the nematode with ever increasing concentrations of glycerol. Rapid methods take advantage of vapor exchange to replace water by ethanol, which in turn is replaced by glycerol. The nematologist will have to experiment with various techniques until a suitable one is found for his specimens.

Slow method, after Golden (8): "Nematodes are left (in fixative) 16 – 24 hr or longer, then two drops of saturated aqueous picric acid are added to the watchglass containing them and it is covered (for example by a closely fitting similar Syracuse watchglass) and placed in an oven at 43°C for 6 – 8 weeks. Formalin – glycerol fixative is added every few days (from a bottle kept in the oven) to replace that lost by evaporation. After 3 – 4 weeks the fluid is allowed to evaporate very slowly until only glycerol remains. Then a few drops of glycerol (from a bottle kept in the oven at 43°C) are added. The nematodes can then be mounted or stored in anhydrous glycerol. Picric acid prevents clearing and fading of the basal portions of nematode stylets; it also prevents growth of molds". This method has the advantage of simplicity and requires no handling of the nematodes. For a more complete set of directions see Ref. (8).

Rapid methods: Seinhorst's *glycerol – ethanol method*, as adapted from Ref. (16): Bring the fixed specimens to a deep dish containing a dilute aqueous solution of ethanol and glycerol (20 parts 96% ethanol + 1 part glycerol + 79 parts distilled water). Place the dish in a desiccator containing 96% ethanol (1/10 the volume of the desiccator). Seal the chamber and incubate at 35 – 40°C for at least 12 hr. The solution in the dish will increase by the deposition of ethanol. Remove some of this addition and replace with a solution of 5% glycerol in 96% ethanol. Place the dish in a Petri dish, partly covered and maintained at 40°C for at least 3 hr. The alcohol will evaporate off, leaving nematodes in pure glycerol. All solutions and dishes must be free of debris.

Still another method, used by A. T. deGrisse, is illustrated in Fig. 4.5. It employs a vacuum desiccator to replace water in nematodes by ethanol.

1. Bring fixed nematodes to a small container with 0.5 – 1 mL fixative to which a small amount of aqueous picric acid has been added.
2. Remove shelf and fill bottom of vacuum desiccator with silica gel, a drying agent. Add enough 95% ethanol to cover silica gel. Replace shelf.
3. Transfer dish with nematodes to desiccator. Cover with wire cage supporting a dish with silica gel.
4. Apply vacuum until the alcohol boils.
5. Seal desiccator and clamp tube from desiccator to vacuum.
6. After a few hours, or overnight in the saturated ethanol vapor, water in nematodes will be replaced by alcohol.
7. Attach a drying tube to the desiccator without breaking the vacuum. Then admit air to the chamber.
8. Add a few drops of anhydrous glycerol to the container with nematodes and incubate at 43°C. Alcohol will evaporate in a few hrs, leaving nematodes in pure glycerol.

Fig. 4.5 Vacuum desiccator to replace water in nematodes by alcohol. (1) Dish with silica gel; (2) shelf; (3) dish with nematodes; (4) wire cage; (5) silica gel covered with 95% ethanol.

D.   *Permanent Mounts on Slides*.   This takes some practice. Transfer several nematodes from the dish of glycerol to a small drop of anhydrous glycerol on a clean glass slide. Try to avoid working in a humid atmosphere. Select three short pieces of fiber glass of approximately the same diameter as that of the nematodes 'and arrange them peripherally in the drop at 120° apart. Add the nematodes, which should be in contact with the slide before adding a cover. Gently heat a round cover glass in a small flame and place it over the drop. The drop should be of a size that just spreads to the edge of the cover glass. Use a micropipette to adjust drop size. Seal the edge of the cover glass with three drops of silicone rubber or nail polish. When these have become firm, ring the cover glass two or three times with a sealant such as Glyceel (G.T. Gurr, Ltd, 136 New King's Rd, London, S.W. 6, England), or Dow Corning's Silastic® or General Electric's Silicone Seal® (4). Note: The silicone rubber should be diluted 30 or 40% with ligroin (naphtha solvent).

E.   *Scanning Electron Microscopy*.   This instrument is rapidly becoming a common adjunct to the light microscope. A beam of electrons is focused on the surface of a specially prepared specimen. Secondary electrons are emitted in such a way that a sharp image of the surface of the object is created. Photographs of nematodes produced by SEM (scanning electron microscopy) are becoming important in several ways in phytonematology: in taxonomy, in studies of host – parasite relations, and to a limited extent in studies of the internal structures of plants and nematodes. Moreover, with additional instrumentation, some chemical analyses of surface materials can be made. The techniques, although very different from those of light microscopy, are accessible to students, and the instrumentation is now available in many laboratories.

Wergin published a thorough account of SEM from which the following is adapted (18). SEM offers five advantages over conventional light microscopy: (a) The images are readily comprehended. (b) The range of magnification extends from 20 – 50,000 ×, and the images are sharp throughout. (c) The depth of field is 300 – 500 times greater than that of the light microscope. (d) The scanning electron microscope can be used for relatively large specimens. (e) Because the stage is easily moved, the object can be viewed from several angles, and three-dimensional images are obtainable.

Mastery of details of the technique requires special instruction. In essence, specimens must be fixed, dried, mounted on a stub, and coated with a fine film of metal before insertion into the microscope for observation. One system of fixation in common use is to subject the specimens to a primary fixative made of 3% glutaraldehyde in 0.05M phosphate buffer, pH 6.8. Fixation proceeds at room temperature for 1 1/2 hr. Special systems have been developed for handling individual phytonematodes to avoid losing the specimens.

Fixed nematodes should be washed in about six complete changes of the above buffer solution, each change lasting 10 – 15 min. Postfixation at room temperature requires a 2-hr immersion in 2% osmium tetroxide in 0.05M phosphate buffer.

For dehydration the organisms are transferred successively through 20, 40, 60, 80, 95, and 100% ethanol during 1 – 2 hr. Following this, three changes of absolute ethanol are required. Following the last ethanol, specimens still in ethanol are transferred to the chamber of a critical- point dryer, which is then sealed. Liquid carbon dioxide replaces the ethanol in a gradual procedure, and the chamber is slowly heated until the critical temperature and pressure are reached. Then the chamber is kept at the critical temperature, pressure is released slowly, and the dehydrated nematodes are ready for mounting on aluminum stubs with special tape.

Finally, the stub is transferred to another instrument, the sputter coater, in which a 200 – 300 ångstrom layer of gold or gold – palladium alloy is deposited over the specimen. The preparation is now ready for examination by SEM.

The section "Technique" in Helminthological Abstracts, Series B, Plant Nematology, contains information on improvements of methods each year.

PREPARATION OF INOCULUM

Investigators requiring populations of nematodes for inoculation of plants or for biochemical studies have several options.

1. Collection of Eggs.   Eggs of *Meloidogyne* spp. can be collected from egg sacs by dissolution of the gelatinous matrices that surround the eggs. Heavily infected roots of favorable hosts bear abundant, exposed egg sacs. To obtain populations of eggs, wash infected roots to remove adhering soil. Then vigorously shake pieces of roots for 4 min in a stoppered flask containing commercial bleach solution diluted fivefold with water (= 1% NaOCl). Pour the fluid over nested sieves and wash to remove hypochlorite. Eggs can be suspended in water and the concentration adjusted for inoculations (11). This method can be combined with differential centrifugation in sucrose solutions to produce concentrated suspensions of eggs.

Eggs may be collected from cysts by maceration. Collect cysts from soil on a 60-mesh screen (250 μm apertures). Wash screenings onto a 100-mesh screen (150 μm apertures) and macerate by gentle rubbing with a rubber stopper to release eggs. Wash sieve over a bucket, using a gentle stream of water, concentrate by centrifugation at 800 g for 5 min in 50-mL conical tubes. A swinging horizontal head rotor is useful. Siphon off the supernatant until about 2 mL remains above the pellet. Stir the pellet and layer the slurry over a sucrose gradient in a 50-mL conical centrifuge tube. The gradient consists of 10 mL each of 50% sucrose overlaid with 40% and 20% sucrose. Centrifuge for 5 min as before. Clean embryonated eggs will concentrate in a band in the 40% layer. In addition the band may contain hatched juveniles and a few fungus spores. Remove the band by siphon and pass through a 63-μm nylon sieve to catch larvae and fungus spores. Eggs are caught on a 31-μm sieve (1).

2. Aseptic Cultivation of Phytonematodes.   Many investigators prepare inocula by cultivation on callus, excised roots, or slices of carrot. Both nematodes and plant tissues must be free of microorganisms. Standard techniques of tissue culture are used to propagate plant roots and callus. Directions for establishing and maintaining nematode cultures are provided in Refs. (6,9,19). Large-scale field experiments are in progress with inoculum from aseptic cultures (15).

REFERENCES

1. Acedo, J. R. and Dropkin, V. H. 1982. Technique for obtaining eggs and juveniles of *Heterodera glycines*. J. Nematol. 14:418 – 420.

2. Byrd, D. W. Jr., Kirkpatrick, T., and Barker, K. R. 1983. An improved technique for clearing and staining plant tissues for detection of nematodes. J. Nematol. 15:142 – 143.

3. Cairns, E. J. 1960. Methods in nematology: A review. In: Nematology, Fundamentals and Recent Advances with Emphasis on Plant Parasitic and Soil Forms (J. N. Sasser and W. R. Jenkins, eds.). Univ. of North Carolina Press, Chapel Hill, Chapter 5, pp. 33 – 84.

4. Caveness, F. 1969. Silicone rubber for ringing cover glasses. J. Nematol 1:96. Note: The silicone rubber should be diluted 30 or 40% with ligroin naphtha solvent.

5. Dropkin, V. H., Smith, W. L., Jr., and Myers, R. F. 1960. Recovery of nematodes from infected roots by maceration. Nematologica 5:285 – 288.

6. Evans, D. A., Sharp, W. R., Amirato, P. V. & Yamada, Y. 1983. Handbook of plant cell cultures. MacMillan, N.Y., 970 pp.

7. Flegg, J. J. M. and Hooper, D. J. 1970. Extraction of free-living stages from soil. In: Laboratory Methods for Work with Plant and Soil Nematodes, 5th ed. (J. F. Southey, ed.). Ministry of Agriculture, Fisheries and Food. H. M. Stationery Office, London, Tech. Bull. 2, pp. 9 – 10.

8. Golden, A. M. 1985. Preparation and mounting nematodes for microscopic observation. In: Plant Nematology Laboratory Manual, (B. M. Zuckerman, W. F. Mai, and M. B. Harrison, eds.). Univ. of Massachusetts Agric. Exp. Stn., Amherst, pp. 189 – 195.

9. Hooper, D. J. 1986. In: Laboratory Methods for work with plant and soil nematodes. 6th ed. (J. F. Southey, ed). Ministry of Agriculture, Fisheries and Food. H. M. Stationery Office, London, Tech. Bull. 2. Available in North America from BERNAN-UNIPUB, 10033-F, M. L. King Highway, Lanham, MD.

10. Hussey, R. S. 1985. Staining nematodes in plant tissue. In: Plant Nematology Laboratory Manual (B. M. Zuckerman, W. F. Mai, and M. B. Harrison, eds). Univ. of Massachusetts Agric. Exp. Station, Amherst, pp. 197 – 199.

11. Hussey, R. S. and Barker, K. R. 1973. A comparison of methods of collecting inocula of *Meloidogyne* spp., including a new technique. Pl. Dis. Reptr. 57:1025 – 1028.

12. Niblack, T. L. and Hussey, R. S. 1985. In: Plant Nematology Laboratory Manual (B. M. Zuckerman, W. F. Mai, and M. B. Harrison, eds.). Univ. of Massachusetts Agric. Exp. Stn., Amherst, pp. 201 – 202.

13. O'Bannon, J. H. and Nyczepir, A. P. 1982. Host range of the Columbia root-knot nematode. Plant Dis. 66:1045 – 1048.

14. Oostenbrink, M. 1960. Estimating nematode populations by some selected methods. In: Nematology, Fundamentals and Recent Advances with Emphasis on Plant Parasitic and Soil Forms (J. N. Sasser and W. R. Jenkins, eds.). Univ. of N.Carolina Press, Chapel Hill, pp. 85 – 102.

15. Rowe, R. C., Riedel, R. M., and Martin, M. J. 1985. Synergistic interactions between *Verticillium dahliae* and *Pratylenchus penetrans* in potato early dying disease. Phytopathology 75:412 – 418.

16. Southey, J.F., ed. 1970. Laboratory Methods for Work with Plant and Soil Nematodes, Tech. Bull. 2, Ministry of Agriculture, Fisheries and Food, H.M.Stationery Office, London.

17. Taylor, A.L. 1971. Introduction to Research on Plant Nematology, an FAO Guide to the Study and Control of Plant-Parasitic Nematodes (rev. ed.). Food and Agriculture Organization of the United Nations, Rome.

18. Wergin, W.P. 1981. Scanning electron microscopic techniques and applications for use in Nematology. In: Plant Parasitic Nematodes, Vol. 3. (B. M. Zuckerman, and R. A. Rohde, eds.). Academic Press, New York, pp.175 – 204.

19. Zuckerman, B.M., Mai, W.F., & Harrison, M.B., eds. 1985. Plant Nematology, Laboratory Manual. Univ. of Massachusetts. Amherst, Massachusetts 01003, pp. 153 – 162.

Huettel, R. N. Carrot disc culture. Exercise 24, 153 – 154

Huettel, R. N. and Rebois, R. V. Culturing plant parasitic nematodes using root explants. Exercise 25, 155 – 158

Riedel, R. M. Establishing *Pratylenchus* on monoxenic culture on alfalfa (*Medicago sativa*) callus tissue. Exercise 26, 159 – 162.

Five

IDENTIFICATION OF PLANT PARASITIC NEMATODES

INTRODUCTION

Classification of organisms into species and genera developed from the need to organize large amounts of information into usable units. For example, investigators in many parts of the world have studied root rots — diseases in which portions of root tissue decay. Nematodes are often found associated with this condition in various plants. Are there any particular kinds of nematodes found in root rots, or is this simply a nonspecific association in which nematodes do not play a significant role? To determine this we must have a system of identification to compare nematodes from tobacco with those from strawberries and other hosts showing similar symptoms.

Differences in structures as seen through the light microscope are used to classify nematodes into genera and species. In general, nematologists attempt to develop classification schemes to represent evolutionary relations of the group. Surface cuticular markings as well as internal structures are important for separating species. Recent technical advances such as scanning electron microscopy are increasing the possibilities of identification beyond those of light microscopy. Biochemical techniques for protein separation are currently under study as tools for identification of nematode species.

Because each nematode taxonomist develops his or her own scheme of classification, the subject is always in a state of flux. As more knowledge develops, prevailing classifications change. This chapter is an introduction to the subject of nematode identification. It will enable you to determine whether a specimen belongs to a family of phytonematodes known to cause important crop damage. With this text alone you cannot identify all the nematodes present in a sample of soil or of plants. But you should be able to decide whether a high population of a known genus of dangerous nematodes is present in a diseased plant or in a sample of soil on which plants grow poorly. The more complete articles and books cited in the references for each genus must be used for accurate nematode identification.

General references listed in Chapter 1 are helpful. Thorne (1) has useful information about most phytonematodes. The Commonwealth Institute of Helminthology (C.I.H.) publishes summaries of information on species of phytonematodes. Each consists of a drawing, measurements, a description of the species, a list of hosts and geographic distribution, information on biology and life history, host – parasite relations, control, and the important literature references. C.I.H. also publishes **Helminthological Abstracts, Series B, Plant Nematology**. Abstracts of the world literature are issued quarterly, together with an annual index. Through this publication one can keep up to date on descriptions of new species and other taxonomic literature. Publications may be obtained from Central Sales, Commonwealth Agricultural Bureaux, Farnham Royal, Slough SL2 3BN, UK. Chapter I on structure and classification of nematodes in Ref. (2) presents an overview of orders and families of plant parasitic and free-living nematodes.

Nematodes in soil samples fall into two categories: (a) Forms without spears may be disregarded—they feed on bacteria or on protozoa. (b) Specimens with spears include plant parasites as well as species that feed on algae, fungi, insects, and other small soil arthropods.

Living anesthetized individuals offer the clearest view of nematode structure. Permanent mounts in glycerin are commonly used for taxonomic descriptions. The diagrams presented here, taken from the literature of nematology, emphasize diagnostic features that distinguish a species or genus from closely related nematodes.

ORDERS AND SUBORDERS

Nematodes form a separate phylum of animals, the phylum Nematoda. More than 10,000 species have been described but these represent only a small portion of the total number of nematode species. In soil, 10 major groups (orders) are present. Two orders contain nematode parasites of plants. The order Tylenchida includes the great majority of plant parasites as well as some parasites of insects and some species that feed on fungi. The order Dorylaimida has a vast array of soil species most of whose life habits are unknown. It includes three genera of plant parasites.

Table 5.1 contains a list of the characteristics of orders of phytonematodes. A representative species of each order is illustrated in Figure 5.1.

The order Tylenchida is divided into two suborders. Members of the

Table 5.1 Characteristics of Orders Containing Phytonematodes

Order Tylenchida	Order Dorylaimida
Narrow, pointed stylet, usually with three posterior knobs to which muscles attach	Most species have a short and broad stylet, with slanted tip, without knobs; or stoma contains teeth
	Phytonematodes have elongated stylets
Pharynx divided into four portions: anterior **procorpus**; enlarged **metacorpus** (= median bulb) with crescentic plates (also called valves) to which attach radial muscles; a narrow **isthmus** encircled by a ring of nerves, and a posterior glandular **bulb** or set of **lobes**	Pharynx without a median bulb; it consists of a narrower anterior and broader posterior portion. It may also be cylindrical throughout
Cuticle has distinct annulations	Annulations of the cuticle are not usually visible when viewed with a light microscope.

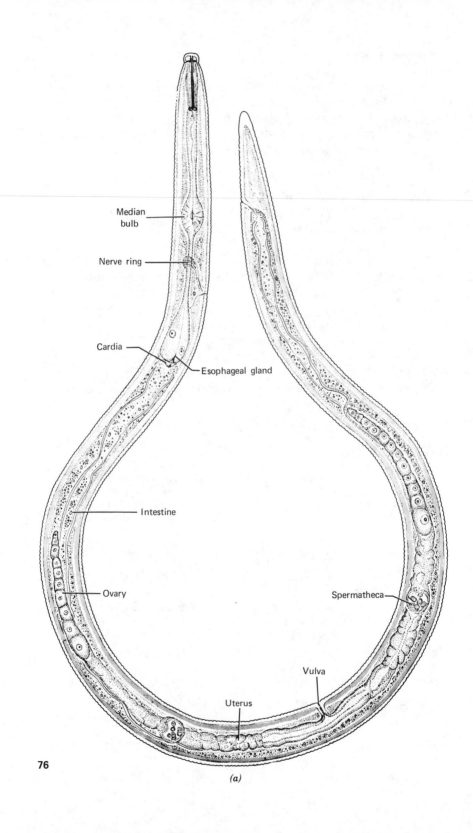

Median
bulb

Nerve ring

Cardia

Esophageal gland

Intestine

Ovary

Spermatheca

Vulva

Uterus

(a)

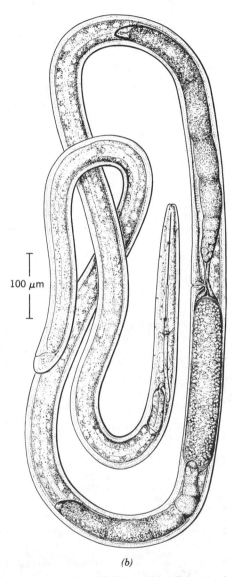

(b)

Fig. 5.1 Orders of phytonematodes. (**a**) Tylenchida, *Tylenchorhynchus cylindricus*, adult female (×540). (From M. R. Siddiqi. 1972. *Tylenchorhynchus cylindricus* C.I.H. Descriptions of Plant-parasitic Nematodes Set 1, No. 7, © Commonwealth Agricultural Bureaux, by permission.); (**b**) Dorylaimida, *Longidorus jonesi*, adult female (×170). (From M. R. Siddiqi. 1972. Studies on the genus *Longidorus* Micoletzky, 1922 (Nematoda: Dorylaimoidea), with descriptions of three new species. Proc. Helminthol. Soc. Wash. 29:177–188, by permission.)

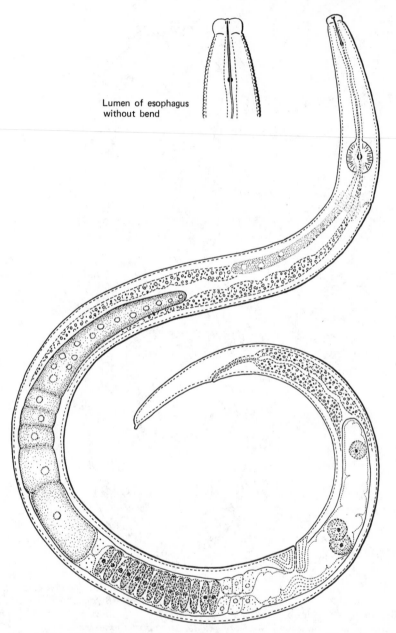

Lumen of esophagus
without bend

Fig. 5.2 *Aphelenchoides composticola*, adult female (×720). (From M. T. Franklin. 1957. *Aphelenchoides composticola*, n. sp. and *A. saprophilus*, n. sp. from mushroom compost and rotting plant tissues. Nematologica 2:306- 313, by permission of E.J. Brill.)

suborder Tylenchina have a stylet with conspicuous knobs, and the duct of the dorsal esophageal gland joins the pharyngeal lumen close to the base of the stylet. The esophageal lumen appears as a canal with a bend at the point of attachment of the gland duct. Nematodes of the suborder Aphelenchina have a stylet with small or no knobs; the dorsal gland duct empties into the metacorpus lumen just anterior to the crescentic plates. The esophageal lumen of the procorpus appears as a straight tube without bends behind the stylet. Figure 5.2 illustrates an example of the suborder Aphelenchina. There are many more species of phytonematodes in the Tylenchina than in the Aphelenchina. Ten families in the Tylenchina contain many species of important plant pests, but only one family of the Aphelenchina has a few species of phytonematodes.

FAMILIES

General characteristics of the families are given. Diagnostic characteristics of each family within the order Tylenchida have been assembled into an illustrated network (lattice) in Table 5.2. Each characteristic listed at the top appears in several ways among the families. For example, the stylet is short or long, or there may be one or two ovaries. To identify the family to which an unknown specimen belongs, take a good look at the specimen, then consult the lattice. Each box has one or more numbers indicating which families listed below have the described form of the character. The lattice may be used in any sequence. For example, nematodes in the family Pratylenchidae resemble those in the family Hoplolaimidae. Both have a strong cephalic framework, either one or two ovaries, and overlapping esophageal glands. But members of the Pratylenchidae have low heads and those in the Hoplolaimidae have more hemispherical heads. The lattice is simply a way of organizing information that will lead you to the correct designation. Once you have assigned the specimen to a family, consult the description and the illustrations of genera within the family to test your identification. See Figs. 5.3–5.10.

Characteristics of families are as follows:

1. Tylenchidae

Cephalic framework absent or weakly developed. Stylet small. Both males and females active, elongate nematodes. Single ovary; vulva between middle of body and anus. Female tail pointed. Males with caudal alae which do not reach to tip of tail. Esophageal glands in a basal bulb that may overlap intestine slightly. *Anguina, Ditylenchus.*

Table 5.2 Lattice for Separation of Families of Order Tylenchida[a]

	Weak	Inter-mediate	Strong	One	Two	Yes	No	Short	Long	Saccate
Procorpus fused with metacorpus						7 – 9	1 – 6 10			
Cephalic framework of females	1, 6	2	3, 4							
Ovary				1, 3, 4 7, 8	2 – 6					
Esophageal glands overlap intestine						3 – 5 10	1, 2 7 – 9			
Stylet								3	5, 7, 8	
Shape of female										6, 9 *Nacobbus*(3), *Rotylenchulus*(4)
Coarse annules						7	8			
Dorsal gland duct in procorpus						1 – 9	10			
Female head low and flattened						3	4			

[a] The numbers 1 – 10 indicate families as follows: 1. Tylenchidae; 2. Tylenchorhynchidae; 3. Pratylenchidae; 4. Hoplolaimidae; 5. Belonolaimidae; 6. Heteroderidae; 7. Criconematidae; 8. Paratylenchidae; 9. Tylenchulidae; 10. Aphelenchoididae.

2. Tylenchorhynchidae

Cephalic framework light to moderate. Stylet well developed with basal knobs. Both sexes active, elongate nematodes from about 0.8 to 1.5 mm long. Two ovaries, vulva in middle of body. Female tail rounded or acute. Caudal alae reach close to tip of tail. Esophageal glands in a basal bulb. *Tylenchorhynchus*.

3. Pratylenchidae

Cephalic framework sclerotized, prominent. Both sexes active, elongate. Head of both sexes low, broad and rounded or flattened anteriorly (except in *Radopholus*), about one half to three- fifths as wide as stylet is long. Strong stylet with large basal knobs. Three esophageal glands in a lobe overlapping intestine. Either one or two ovaries. Female tail twice or more as long as body width at anus. Caudal alae reach to tail tip. *Pratylenchus, Radopholus, Hirschmanniella, Nacobbus*.

4. Hoplolaimidae

Head high, convex – conoid or broadly rounded. Both sexes active, elongate. Cephalic framework well developed. Thick stylet with prominent basal knobs. Esophageal glands overlap intestine. Annules of cuticle prominent. One or two ovaries. Female tail short, usually less than two anal body widths long. Caudal alae extend to tip of tail. *Hoplolaimus, Scutellonema, Rotylenchus, Helicotylenchus, Rotylenchulus*.

5. Belonolaimidae

Both males and females elongate, with well marked annules. Head offset, cephalic framework moderate. Stylet slender, long. Esophageal glands in lobe overlapping intestine. Two ovaries. Female tail rounded, at least twice as long or longer than body width at anus. Caudal alae terminal. *Belonolaimus*.

6. Heteroderidae

Pronounced difference between sexes. Females swollen, pear-shaped, globular, or lemon-shaped; sessile in or on roots. Males elongate, vermiform, active. Cephalic framework of females weak, not sclerotized and more developed in males. No caudal alae. In *Heterodera* and *Globodera*, females become cysts and in *Meloidogyne* they remain soft. Two ovaries, vulva terminal or behind middle of body. *Globodera, Heterodera, Meloidogyne* are the major genera of economic importance.

7. Criconematidae

Cuticle generally with coarse annulation; in some species annules have overlapping scales. Large oval metacorpus broadly fused with procorpus into a single subdivision of the pharynx. Crescentic plates elongate. Isthmus short and narrow, glands in a small basal bulb. Vulva posterior, single ovary. Males absent or degenerate. Stylet of female very long in some species. *Criconemoides, Macroposthonia, Criconema, Hemicycliophora.*

8. Paratylenchidae

Small nematodes with pharynx organized as in Criconematidae. Body annules fine, not ornamented or overlapping. One ovary, vulva posterior. Female stylet well developed; male stylet absent or reduced. *Paratylenchus.*

9. Tylenchulidae

Female saccate or subspherical. Pharynx organized as in Criconematidae. Stylet short, well developed in females and reduced or absent in males. Adult females are partly buried in root, with their swollen posteriors protruding from the surface. *Tylenchulus.*

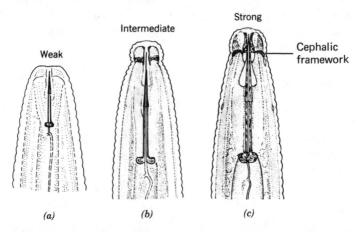

Fig. 5.3 Anterior portion of three nematodes to illustrate cephalic framework. **(a)** *Ditylenchus myceliophagus* (× 1933), weak cephalic framework (From J. J. Hesling. **(b)** *Tylenchorhynchus cylindricus* (× 1150), intermediate (From M. R. Siddiqi. 1972 *Tylenchorhynchus cylindricus*. C.I.H. Descriptions of Plant-parasitic Nematodes Set 1, No. 7, © Commonwealth Agricultural Bureaux, by permission.); **(c)** *Rotylenchus robustus* (× 890), strong. (From S. A. Sher. 1965. Revision of the Hoplolaiminae (Nematoda) V. *Rotylenchus* (Filipjev, 1936). Nematologica 11:173–198, by permission of E.J.Brill.)

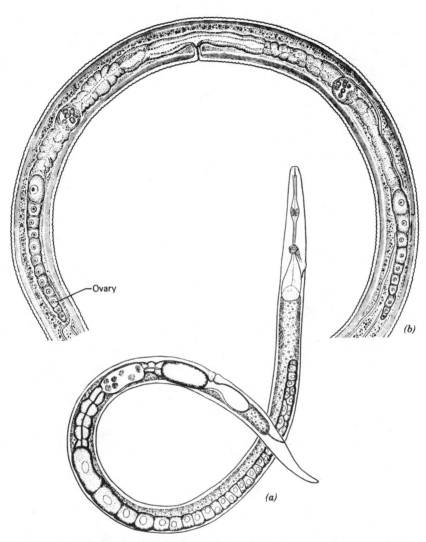

Fig. 5.4 One and two ovaries. (**a**) one ovary – *Ditylenchus myceliophagus* (×206). (From J. J. Hesling. 1974. *Ditylenchus myceliophagus*. C.I.H. Descriptions of Plant-parasitic Nematodes, Set 3, No. 36, © Commonwealth Agricultural Bureaux, by permission.); (**b**) two ovaries – *Tylenchorhynchus cylindricus* (×520) (From M. R. Siddiqi. 1972. *Tylenchorhynchus cylindricus*. C.I.H. Descriptions of Plant-parasitic Nematodes Set 1, No. 7, © Commonwealth Agricultural Bureaux, by permission.)

10.   Aphelenchoididae

Dorsal esophageal gland opens into metacorpus anterior to crescentic plates. Metacorpus usually large. Stylet without conspicuous knobs. One ovary. Spicules thorn-shaped. Lateral field with less than six incisures. *Aphelenchoides, Bursaphelenchus, Rhadinaphelenchus.*

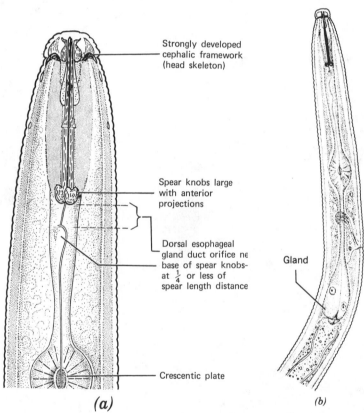

Strongly developed
cephalic framework
(head skeleton)

Spear knobs large
with anterior
projections

Dorsal esophageal
gland duct orifice ne
base of spear knobs-
at $\frac{1}{4}$ or less of
spear length distance

Gland

Crescentic plate

(a)   *(b)*

Fig. 5.5   Position of esophageal glands. (a) Overlapping intestine— *Hoplolaimus stephanus* (×747). (From S. A. Sher. 1963. Revision of the Hoplolaiminae (Nematoda). II. *Hoplolaimus* Daday 1905 and *Aorolaimus* n. gen. Nematologica 9:267–895., by permission of E.J.Brill.); (b) not overlapping intestine—*Tylenchorhynchus cylindricus* (×750). (From M. R. Siddiqi. 1972. *Tylenchorhynchus cylindricus.* C.I.H. Descriptions of Plant-parasitic Nematodes Set 1, No. 7, © Commonwealth Agricultural Bureaux, by permission.)

11. Longidoridae

Long, slender nematodes with very elongate stylets. Pharynx with long anterior muscular portion and a short posterior glandular portion of larger diameter. *Longidorus, Xiphinema.*

12. Trichodoridae

Short, broad nematodes with thick cuticle and curved, three-part stylet. Tail blunt and round. Esophageal glands form a basal bulb. *Trichodorus, Paratrichodorus.*

Figures 5.3 – 5.10 illustrate characters for identification of families as listed in the lattice.

Major genera of phytoparasitic nematodes of economic importance are arranged here according to taxonomic sequence (2). They are presented in this order in Chapter 6.

Order Tylenchida
  Suborder Tylenchina
    Superfamily Tylenchoidea
      Family Tylenchidae: *Anguina, Ditylenchus*

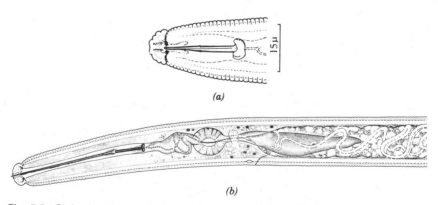

(a)

(b)

Fig. 5.6 Stylet length. (a) Short stylet—head of 2nd stage juvenile of *Heterodera schachtii* (×850). (From M. T. Franklin. 1972. *Heterodera schachtii,* C.I.H. Descriptions of Plant-parasitic Nematodes, Set 1, No. 1, © Commonwealth Agricultural Bureaux, by permission.); (b) long stylet—head of *Belonolaimus longicaudatus* (×367). (From K. J. Orton Williams. 1974. *Belonolaimus longicaudatus,* C.I.H. Descriptions of Plant-parasitic Nematodes, Set 3, No. 40, © Commonwealth Agricultural Bureaux, by permission.)

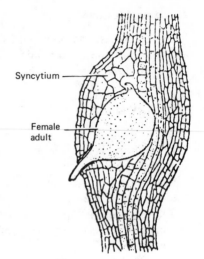

Fig. 5.7 Shape of female—saccate—*Nacobbus dorsalis* (× 20). (From J. B. Goodey. 1963. Soil and Freshwater Nematodes. Wiley, New York, by permission.)

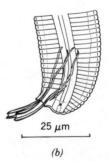

25 μm

(a)   *(b)*

Fig. 5.8 Character of annulation. **(a)** Coarse annules—*Macroposthonia xenoplax* (× 390). (From K. J. Orton Williams. 1972. *Macroposthonia xenoplax*. C.I.H. Descriptions of Plant-parasitic Nematodes, Set 1, No. 12, © Commonwealth Agricultural Bureaux, by permission.); **(b)** not coarse annules—*Heterodera schachtii* (× 744). (From M. T. Franklin. 1972. *Heterodera schachtii*. C.I.H. Descriptions of Plant-parasitic Nematodes, Set 1, No. 1., © Commonwealth Agricultural Bureaux, by permission.)

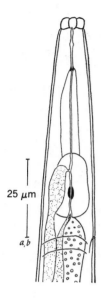

Fig. 5.9 Position of dorsal gland orifice not in procorpus, but in the median bulb—*Cryptaphelenchoides macrobulbosus*. Compare Fig. 6.12, showing the dorsal gland orifice in the procorpus. (From J. B. Goodey. 1963. Soil and Freshwater Nematodes. Wiley, New York, by permission.)

Family Tylenchorhynchidae: *Tylenchorhynchus*
Family Pratylenchidae: *Pratylenchus, Radopholus,*
          Hirschmaniella, Nacobbus
Family Hoplolaimidae: *Hoplolaimus, Scutellonema, Rotylenchus,*
          Helicotylenchus, Rotylenchulus
Family Belonolaimidae: *Belonolaimus*

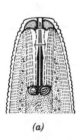

(a)

(b)

Fig. 5.10 Shape of head. (a) Female head flattened—*Pratylenchus*. (From G. Thorne. 1961. Principles of Nematology, McGraw Hill, New York, by permission.); (b) rounded—*Rotylenchus robustus*. (From S. A. Sher. 1965. Revision of the Hoplolaiminae (Nematoda) V. *Rotylenchus* (Filipjev, 1936). Nematologica 11:173 – 198, by permission of E.J.Brill.)

Superfamily Heteroidea
  Family Heteroderidae: *Heterodera, Globodera, Meloidogyne*
Superfamily Criconematoidea
  Family Criconematidae: *Criconemoides, Macroposthonia,*
                         Criconema, Hemicycliophora
  Family Paratylenchidae: *Paratylenchus*
  Family Tylenchulidae: *Tylenchulus*
  Suborder Aphelenchina
  Superfamily Aphelenchoidea
  Family Aphelenchoididae: *Aphelenchoides, Bursaphelenchus,*
                           Rhadinaphelenchus
Order Dorylaimida
  Family Longidoridae: *Longidorus, Xiphinema*
  Family Trichodoridae: *Trichodorus, Paratrichodorus*

REFERENCES

1. Thorne, G. 1961. Principles of Nematology. McGraw-Hill, New York.
2. Hooper, D. J. 1978. Structure and Classification of Nematodes. In: Plant Nematology, Southey, J.F., ed. GD1, Ministry of Agriculture, Fisheries and Food, H.M. Stationery Office, London, pp. 3 – 45.

Six

THE GENERA OF PHYTONEMATODES

INTRODUCTION

Twenty-nine important genera are described in order of their taxonomic relationships as listed in Chapter 5. For each genus the text describes classification, morphology, biology, pathology induced in host plants, interaction with other pathogens, and control. Recent research is briefly reviewed and a list of references concludes the discussion. Information necessary for identification of each genus is included. The discussion of research was prepared by scanning **Helminthological Abstracts, Series B**, for 1981 – 1985. These abstracts represent the entire world literature and provide an overview of current research on each genus. Research is presented to stimulate deeper interest in phytonematology. From such interest and research comes improvement of the practical management of agricultural problems in which nematodes have a role.

ANGUINA

Introduction

This genus includes gall-forming nematodes that inhabit aerial parts of plants. They are important on wheat, especially in countries where seed cleaning does not eliminate infected seed. They are also associated with toxin-producing bacteria on certain grasses serving as food for livestock.

Classification

Order Tylenchida; suborder Tylenchina; superfamily Tylenchoidea; family Tylenchidae; subfamily Anguininae.

Morphology

Sexes are separate. **Females** are obese, tapered toward both ends, reaching lengths of up to 5 mm in some species but 1.5 mm or more in others. When heated they coil in a spiral with the ventral surface inward. Ratio of overall length to greatest width is 20 or less in some species, but others are more slender. Head is not offset, lips are low and flattened, labial skeleton is weak. Stylet is small, around 10 μm long, with small knobs. Three large esophageal glands usually overlap the intestine in a basal bulb in most but not all species. The vulva is posterior, close to 90% of the distance from lips to tail. The tail constitutes from 1/30 to 1/50 of total body length. Female gonad is reflexed with multiple rows of oocytes. There is a postvulval sac. **Males** are smaller than females and more slender. The testis is reflexed and has multiple rows of spermatocytes. The bursa extends from a position just anterior to the spicules and ends just short of the tail tip. Annulations in both sexes are fine and difficult to see except in the esophageal region. Lateral lines are inconspicuous (7). Figure 6.1 is a drawing of *Anguina tritici*, the wheat gall nematode.

Biology

Infective juveniles invade young plants and induce galls within which nematodes mature to adult males and females, usually a few in each gall. Very many eggs are deposited from which juveniles emerge within the gall. Development proceeds through four stages; the first molt may occur in the egg. In some species only one generation develops within a gall during the

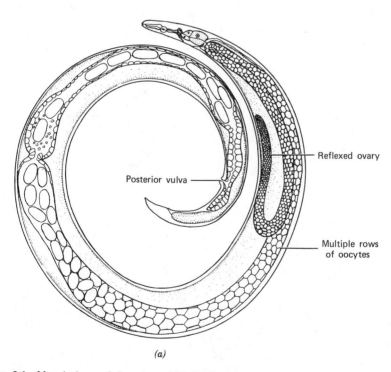

(a)

Fig. 6.1 Morphology of *Anguina tritici*. (a) Female coiled ventrally (×84); (b) Male (×84); (c) Male tail (×540); (d) Female esophageal region (×424). (From J. F. Southey. 1972 C.I.H. Descriptions of Plant- parasitic Nematodes Set 1, No. 13, © Commonwealth Agricultural Bureaux, by permission.)

Reflexed testis

(b)

Fig. 6.1 (*Continued*)

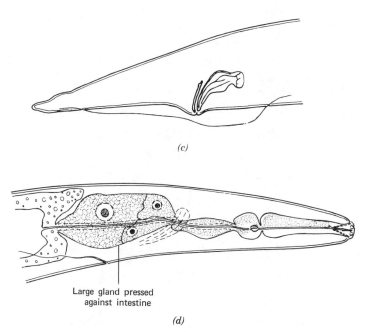

(c)

Large gland pressed
against intestine

(d)

Fig. 6.1 (*Continued*)

growing season. In others there may be more than one. Juveniles survive
for extended periods (years!) in galls under dry conditions; metabolism is
reduced to low levels and the parasites survive the seasons between growing
crops. When moisture returns, metabolism resumes (1).

Pathology

All species of *Anguina* induce galls, some of which are of characteristic
size, shape, and location. Figure 6.2 illustrates leaf galls. Infective stages
(second or third, depending on the species) invade rapidly growing tissues
of seedlings and are carried up as the plant grows. Galls develop by enhanced multiplication and enlargement of parenchyma and vascular cells.
Mesophyll cells of leaves into which infective stages have entered develop
dense cytoplasm and enlarged nuclei and may become multinucleate. Cells
separate in the center of a gall, forming a cavity. Chloroplasts may be
lacking in a tissue where active photosynthesis normally takes place. A
gall is the site of very active metabolism and enhanced accumulation of
nutrients. Both increasing parasite numbers and the abnormal growth re-

Fig. 6.2 Symptoms of *Anguina moxae* on *Artemisia asiatica* (From Y. E. Choi and P. A. A. Loof. 1973. Redescription of *Anguina moxae* Yokoo & Choi, 1968 (Tylenchina). Nematologica 19:285 – 292, by permission of E.J. Brill.)

sponse stress the host. Malformed leaves contribute subnormal amounts of photosynthate and flower galls lose their ability to form seeds (7). Although we do not know the nature of esophageal gland contents of *Anguina* spp. we may suppose that these nematodes inject or excrete auxins or similar growth regulators into plants.

Leaves of infected wheat are small and twisted. A flower gall may have up to 40 adults of each sex and contain 30,000 or more eggs and juveniles. Infected flowers develop into small, dark galls that mix with the seed during harvest or fall to the ground (Fig. 6.3).

Interaction with Other Pathogens

Two important problems result from an association of *Anguina* spp. with *Corynebacterium* spp. In India there are two distinct symptoms in wheat

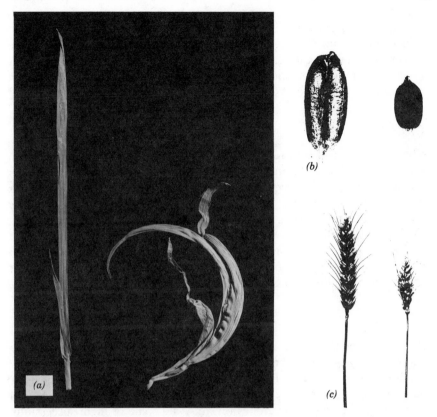

Fig. 6.3 Symptoms on wheat caused by *Anguina tritici*. (a) Normal leaf, (left); twisted emerging leaf (right); (b) normal seed (left), galled seed (right); (c) normal head (left), diseased head (right). (From P. S. Lehman. 1979. Seed and Leaf Gall Nematodes of the genus *Anguina* occurring in North America. Nematol. Circ. No. 55, Div. Plant Ind., Florida Dept. of Agriculture and Consumer Services, Gainesville, by permission.)

infected with *A. tritici*. Plants harboring the nematode, but not the bacteria, show typical ear-cockle in which wheat grains are replaced by dark galls and the ear is small and malformed. The ears of plants with both the nematode and bacteria exude a slimy, yellow ooze. This condition is called yellow ear rot or "tundu." Inoculation of the bacteria alone causes no symptoms (2).

A similar association occurs in Australia where annual ryegrass with both *A. agrostis* and *Corynebacterium rathayi* becomes toxic to domestic animals. Apparently bacteria develop only in galls induced by nematodes. A neurotoxin that kills sheep and other large mammals accumulates when galls mature late in the season. Toxin appears to be concentrated in walls of galls and may be a plant product in response to bacteria (8). Even short periods of exposure to toxic ryegrass can be fatal to sheep.

Control

The aim of control is to reduce the number of infected galls in a crop. In the past *A. tritici* was a problem on wheat in many parts of the world but modern seed cleaning methods now control these nematodes in major wheat producing regions. Rotation or moist fallow for a year and heavy grazing during the growing season may prevent production of infected grass seed. Cleaning eliminates the sowing of galls and herbicides may be used to abort flower production. Seed treatment with methyl bromide is also effective but laborious. Most species of *Anguina* and related genera do not affect economically important plants; therefore control is not attempted.

Research

Nematodes of the genus *Anguina* have attracted researchers for many years. Ability of these nematodes to survive prolonged storage under dry conditions is spectacular. Limber (3) opened dried galls after storage for 32 years under low relative humidity in sealed glass vials. When such galls were placed in water, nematodes revived, and some of them successfully invaded wheat seedlings.

Ability of nematodes to survive freezing is also of interest. Lipids from *A. tritici* and *Meloidogyne hapla* withstood temperatures from near 0 to 30°C without change of their physical state, whereas lipids from *Caenorhabditis elegans* and *M. javanica* changed from a liquid state above 10°C to a liquid-crystalline state below this temperature (4). Such a fundamental difference in lipids may explain the ability of the former two nematodes to maintain metabolic activity at cool temperatures in contrast to the latter. The authors suggest that lipids of the mitochondrial membranes are critical in this connection.

Ready availability of large numbers of nematodes from galls enabled investigators to study the attraction of nematodes to the cation in an electric field (9). In an extensive series of trials the authors determined optimal concentrations of various salts for maximum attraction to the cathode. They applied a magnetic field at right angles to the electrical field and found that the ability of the nematodes to orient was destroyed if the salts became oriented in the magnetic field, but not if salts were unaffected. They proposed that nerves in the amphids can sense a change in potential resulting from a disturbance of the ionic atmosphere within the amphid. If true, this may be related to the observation that growing roots have nonuniform electric potentials along their length and that nematodes often enter roots in certain zones.

Nothanguina phyllobia is common on silverleaf nightshade (*Solanum elaeagnifolium*) in certain fields in Texas. Seed production by infected plants is sharply reduced. Some authors suggest that this nematode has potential for biological control of a noxious weed (5). Another report describes the histopathology of this infection (6).

References

1. Bhatt, B. D. and Rohde, R .A. 1970. The influence of environmental factors on the respiration of plant-parasitic nematodes. J. Nematol. 2:277 – 285.

2. Gupta, P. and Swarup, G. 1972. Ear-cockle and yellow ear-rot diseases of wheat. II. Nematode bacterial association. Nematologica 18:320 – 324.

3. Limber, D. P. 1973. Notes on the longevity of *Anguina tritici* (Steinbuch,1799) Filipjev, 1936, and its ability to invade wheat seedlings after thirty-two years of dormancy. Proc. Helminthol. Soc. Wash. 40:272 – 274.

4. Lyons, J. M., Keith, A. D., and Thomason, I. J. 1975. Temperature-induced phase transitions in nematode lipids and their influence on respiration. J. Nematol. 7:98 – 104.

5. Orr, C. C., Abernathy, J. R., and Hudspeth, E. B. 1975. *Nothanguina phyllobia*, a nematode parasite of silverleaf nightshade. Plant Dis. Rep. 59:416 – 418.

6. Skinner, J. A., Orr, C. C., and Robinson, A. F. 1980. Histopathogenesis of the galls induced by *Nothanguina phyllobia* in *Solanum elaeagnifolium*. J. Nematol. 12:141 – 150.

7. Southey, J. F. 1972. Commonwealth Institute of Helminthology, Descriptions of Plant-parasitic Nematodes, Set 1 No. 13: *Anguina tritici*; see also Set 2, No. 20: *A. agrostis*; Set 4, No. 53: *A.graminis*; see also Southey, J. F., ed. 1978. Plant Nematology. Ministry of Agriculture, Fisheries, and Foods, GD$_1$. London, H. M. Stationery Office. 440 pp.

8. Stynes, B. A., Peterson, D. S., Lloyd, J., Payne, A. L., and Lanigan, G.W. 1979. The production of toxin in annual ryegrass, *Lolium rigidum* infected with a nematode, *Anguina* sp. and *Corynebacterium rathayi*. Aust. J. Agric. Res. 30:201 – 209.

9. Sukul, N. C., Das, P. K., and Ghosh, S. K. 1975. Cation-mediated orientation of nematodes under electrical fields. Nematologica 21:145 – 150.

DITYLENCHUS

Introduction

This genus contains a variety of species, some of which feed on fungi as well as plant tissues. Pathogenic species attack roots and aerial parts of

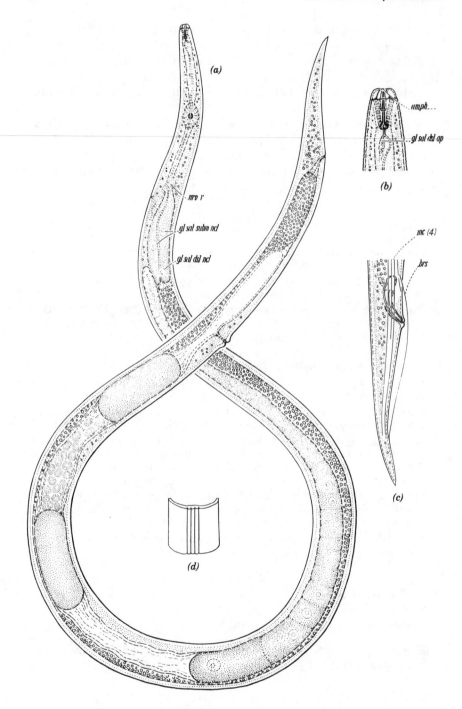

plants. The bulb and stem nematode *D. dipsaci* is a serious pest of many crops and was one of the earliest phytonematodes to be studied intensely. *D. destructor* is a pest of potato and *D. myceliophagus* affects mushroom production. The genus *Ditylenchus* includes two other species of some importance. *D. radicicolus* induces galls on roots of grasses, including barley and rye, annual meadow grass (*Poa annua*), and other grasses. *D. angustus* causes a serious disease of rice ("ufra") in Bangladesh and Thailand.

Classification

Order Tylenchida; suborder Tylenchina; superfamily Tylenchoidea; family Tylenchidae; subfamily Anguininae.

Morphology

These are elongate nematodes, ranging from 0.6 to 1.5 mm long. Sexes are separate and generally similar. The species have low and flat lips, slightly or not offset from the body, a delicate stylet, usually under 15μm long, and weak cephalic sclerotization. A fusiform median bulb is located anterior to the esophageal midpoint. Esophageal glands form a club-shaped or spoon-shaped basal bulb abutting or slightly overlapping the intestine. The vulva is in the posterior quarter of the body; the single gonad has a postvulval sac extending about halfway from vulva to anus. Cuticle annulations are fine; lateral fields have four or six incisures. The male bursa extends part of the distance to the tail tip (9). Figure 6.4 depicts the morphology of *D. dipsaci*.

Biology

Life cycles of *Ditylenchus* spp. are simple. *D. myceliophagus* takes 18 days to complete a generation at 23°C and longer at lower temperatures. *D. dipsaci* is adapted to cooler temperatures, requiring 21 days at 15°C. It invades all aboveground tissues of its host and is present in the seed at harvest.

Fig. 6.4 Morphology of *Ditylenchus dipsaci*. (a) Adult female ×330; (b) female head ×1000; (c) male tail ×500; (d) lateral incisures ×1000. (a) From G. Thorne. 1961. Principles of Nematology. McGraw-Hill, New York, by permission; (b – d) From D. J. Hooper. 1972. Descriptions of Plant-parasitic Nematodes. Set 1, No. 14 © Commonwealth Agricultural Bureaux, by permission. Abbreviations for (a): gl sal dsl ncl, nucleus of dorsal esophageal gland; gl sal subm ncl, nucleus of subventral esophageal gland; nrv r, nerve ring; and for (b) and (c): amph, amphid; gl sal dsl op, opening of dorsal esophageal gland; inc, incisure; brs, bursa.

Southey (12) offers the following information for differentiating British species of *Ditylenchus:*

Species	Part of Host	Tail Tip	Lateral Incisures	Length of Postvulval Sac Relative to Vulva – Anus distance
D. dipsaci	Leaves, stems, and flowers	Sharply pointed	4	$\frac{1}{2}$
D. destructor	Underground parts of plants, especially potato tubers	Rounded	6	$\frac{2}{3}$
D. myceliophagus	Mushroom mycelium	Rounded	6	$\frac{1}{2}$

Pathology

D. dipsaci induces profound changes in infected plants. Such plants are stunted, stems are swollen, often twisted (Fig. 6.5a). Cells around the nematodes dissolve and tissues become fragile, subject to desiccation (Fig. 6.5b) (14). Alfalfa fields with high populations of this nematode may appear white as the tissues dry. Leaves of gladiolus develop characteristic galls ("spikkels"). Strawberries have swollen leaves and petioles. Leaf bases of oats swell, giving a characteristic appearance ("tuliproot"). Flower bulbs are attacked by the nematodes and by associated microorganisms so that individual scales rot and appear as brown rings when the bulbs are cut in cross section (Fig. 6.5c). When a bulb is sufficiently rotted, nematodes

Fig. 6.5 Symptoms of damage by *Ditylenchus* spp. **(a)** Clover, *D. dipsaci*, left—two diseased seedlings, showing stem thickening and dwarfing; right—two healthy seedlings. **(b)** Gall of *Cirsium arvense*, longitudinal section × 10 of stem taken below the apex (A) inhabited by *D. dipsaci*. Note the central cavity (cc) and nematodes (arrows). **(c)** Longitudinal section through a hyacinth bulb with necrotic areas, *D. dipsaci*. **(d)** Rice affected by *D. angustus*; left healthy; right, infected. **(a)** and **(c)** From I. N. Filipjev and J.H. Schuurmans Stekhoven. 1941. A Manual of Agricultural Helminthology, 878 pp., E.J. Brill, Leiden, by permission. **(b)** From A. K. Watson and J. D. Shorthouse. 1979. Gall formation in *Cirsium arvense* by *Ditylenchus dipsaci*. J. Nematol. 11:16 – 22.; **(d)** From S. H. Ou. 1976. Rice Diseases. 368 pp. Commonwealth Mycological Institute, Kew, G.B., © Commonwealth Agricultural Bureaux, by permission.

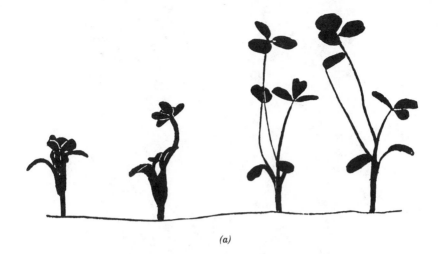

(a)

(b)

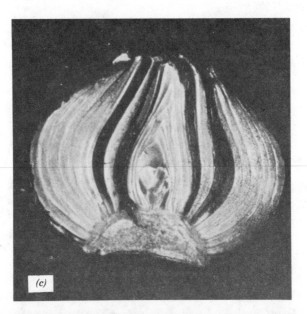

Fig. 6.5 (*Continued*)

leave and accumulate at the base in masses of J–4s known as "eelworm wool."

D. destructor enters potato tubers through lenticels and small, white spots develop just below the surface. As the nematodes multiply, infested areas coalesce into lesions of dry, granular tissue. Secondary invasion by other organisms results in a generalized dark brown rot. This nematode is a danger to potatoes in storage because infected tubers rot and the infection spreads to adjacent tubers.

D. myceliophagus destroys mushroom mycelium and depresses the yield of mushrooms. Very large populations that aggregate into clumps often develop. Nematodes in such clumps are resistant to drying; insects carry them to uninfected mushroom beds.

D. angustus feeds as an ectoparasite on growing, tender tissues and destroys developing leaves and flowers (Fig. 6.5**d**). The nematodes survive dry conditions in a coiled, dormant state.

Interaction with Other Pathogens

Lesions induced by *D. destructor* are subject to invasion by fungi, bacteria, and mites. The nematodes feed on fungi as well as on the plant so that this interaction is not the usual one resulting in enhanced pathogenicity. *D. dipsaci* is believed to increase the severity of bacterial wilt (caused by *Corynebacterium insidiosum*) of some alfalfa varieties (7).

Control

The flower bulb industry suffered great losses from bulb and stem nematodes until hot water methods of control were developed. Resarch on precise control of temperature, time, and added fungicides produced practical control of *D. dipsaci*. To kill the nematodes without harming the plant, dormant narcissus bulbs are soaked in water containing fungicides and maintained at 44 – 45°C for three hours. Similar treatment also serves for onions, garlic, and shallots. Strict sanitation enforced by a strong regulatory program of quarantines is currently practiced in the flower industry. Fumigation of infested seed with methyl bromide prevents subsequent infection with seedborne nematodes (6). Genetic control is also useful. Plant breeders have produced cultivars of alfalfa, oats, and rye resistant to the bulb and stem nematode.

The wide host range of *D. destructor* and its ability to feed on fungi make control difficult. Control requires use of nematode-free seed potatoes, strict quarantine on the movement of infested potatoes, and nematicide treatment of soil. Weed hosts can maintain the nematodes when potatoes are not planted.

Control of *D. myceliophagus* in infested mushroom beds is impractical without destroying the crop. Heat (60°C throughout the compost and casing) will kill nematodes; and certain nematicides, especially thionazin, may prove useful. But above all, sanitation and insect control are essential for successful mushroom production.

Research

Ditylenchus dipsaci was one of the earliest nematodes to be studied critically; research on this species has contributed greatly to the development of phytonematology. Need for resistant cultivars led to advances in large-scale production and handling of nematode inoculum. Callus cultures for production of inoculum and for maintenance of separate races have helped in this respect (1). The striking pathology stimulated studies of enzymes from nematodes to understand pathogenesis. And the puzzling observations that some soils are hazardous for bulb and stem disease whereas others are not led to careful investigation of the ecology of the organisms. The existence of distinct races differing in host range stimulated attention to genetics of pathogenicity. A few key papers from the literature are cited here.

Ditylenchus dipsaci causes host cells to separate and disintegrate. Aqueous extracts of aseptic nematodes contain two pectinases, one of which actively macerates plant tissues. Infected callus tissues also have pectinases. Upon removal of nematodes, enzyme activity falls, and when nematodes are returned to the tissues, pectinase activity rises (10). We may postulate, therefore, that the parasites produce pectinases and that these probably initiate the observed tissue destruction.

Disturbances of growth are common in infections of plants with phytonematodes. Alfalfa plants infected with *D. dipsaci* are stunted. A hypothesis that stunting results from inadequate concentrations of growth regulator in the shoot tip was tested (13). During incubation in water, the nematode releases a heat-labile auxin-inactivating system, probably containing one or more enzymes. On the other hand, stem and leaf galls and twisted growth associated with this nematode may be signs of growth regulator imbalance rather than inadequacy. A thorough study was made of growth regulator activity in extracts of *D. dipsaci* and of a species that feeds on fungi (*D. triformis*). The authors found a methyl ester of IAA in both nematodes, but the growth promoting activity of extracts from *D. dipsaci* was higher than that from *D. triformis* and absent from the vinegar eel (*Turbatrix aceti*) (4). Further analysis of the relation between phytonematodes and plant hormones will lead to greater understanding of pathogenesis.

Soil harbors many organisms in addition to nematodes. It is always hard to separate effects of phytonematodes on plants from those of associated microorganisms. An interaction between *D. dipsaci* and *Pseudomonas fluorescens* bacteria has been found in a disease of garlic. The authors suggest that the nematodes transport the bacteria into plant parenchyma tissues (3).

Development of resistant cultivars is a laborious process requiring large amounts of nematode inoculum. Because the species *D. dipsaci* contains populations of different host ranges, breeders must use nematodes of known host ranges. Callus cultures have provided inoculum for plant breeders in Sweden and elsewhere for many years. Red clover and lucerne populations of *D. dipsaci* that had been on callus cultures for 14 years showed no changes in host specificity or in virulence of the nematodes (1). A collection of various populations maintained on callus cultures has been used to study genetics of host specificity (5). Callus cultures are also useful for work with several species of phytonematodes in addition to *D. dipsaci*.

Maize is a crop of increasing importance in Europe. *D. dipsaci* induces a condition called "early toppling" in which corn plants fall over before reaching maturity. Early plantings of corn (April 9) in France showed severe damage by late June but sowing in May greatly reduced the hazard (2). Cold and wet weather after sowing increases risk even with low soil populations of the nematode whereas warm and dry weather protects plants even from high soil populations (8). It is not yet clear whether the effect of planting date or of soil conditions is on the nematodes in soil or on the interaction of nematodes and plants.

Biochemists have tried to understand how some species of nematodes resist drying but others do not. Resistant nematodes produce protective compounds during desiccation (15). Techniques for freezing *D. dipsaci* without killing the nematodes have been perfected. A proportion of the nematodes survived storage in liquid nitrogen for at least 18 months (11). This technique will be important for preservation of genetic stocks in the future as work on genetics of phytonematodes expands. Similar methods are already in use for preservation of mutant strains of *Caenorhabditis*.

References

1. Bingefors, Sv. and Bingefors, Si. 1976. Rearing stem nematode inoculum for breeding purposes. Swed. J. Agri. Res. 6:13 – 17.
2. Caubel, G. 1973. Effect of sowing date on the early toppling of maize due to the stem nematode (*Ditylenchus dipsaci*). Sci. Agron. Rennes:101 – 107.

3. Caubel, G. and Samson, R. 1984. Effect of the stem nematode, *Ditylenchus dipsaci*, on the development of "café au lait" bacteriosis in garlic (*Allium sativum*) caused by *Pseudomonas fluorescens*. Agronomie 4: 311 – 313.

4. Cutler, H. G. and Krusberg, L. R. 1968. Plant growth regulators in *Ditylenchus dipsaci, Ditylenchus triformis* and host tissues. Plant Cell Physiol. 9:479 – 497.

5. Eriksson, K. B. 1974. Intraspecific variation in *Ditylenchus dipsaci*. I. Compatibility tests with races. Nematologica 20:147 – 162.

6. Hague, N. G. M. 1968. Fumigation of shallot seed with methyl bromide to control stem eelworm, *Ditylenchus dipsaci*. Plant Pathol. 17: 127 – 128.

7. Hawn, E. J. 1970. New technique for studying a bacterium/nematode interaction in alfalfa. J. Nematology 2:272 – 273.

8. Hirling, W. 1974. Pathogenic nematodes on maize in Baden- Wurttemberg. I. "Toppling disease" of maize due to *Ditylenchus dipsaci*. Anz. Schaedlingskd, Pflanz.-Umweltschutz 47:33 – 39.

9. Hooper, D. J. 1972. Commonwealth Institute of Helminthology, Descriptions of Plant-parasitic Nematodes, Set 1, No. 14 *Ditylenchus dipsaci*; see also Set 3, No. 36: *D.destructor*; Set 5, No. 64: *D. myceliophagus*.

10. Riedel, R. M. and Mai, W. F., 1971. Pectinases in aqueous extracts of *Ditylenchus dipsaci*. J. Nematol. 3:28 – 38.

11. Sayre, R. M. and Hwang, S. W. 1975. Freezing and storing *Ditylenchus dipsaci* in liquid nitrogen. Jour. of Nematology 7:199 – 202.

12. Southey, J. F., ed. 1978. Plant Nematology. Ministry of Agriculture, Fisheries and Food, GD$_1$. H. M. Stationery Office, London, 440 pp.

13. Viglierchio, D. R. and Yu, P. K. 1965. Plant parasitic nematodes: a new mechanism for injury of hosts. Science 147:1301 – 1303.

14. Watson, A. K. and Shorthouse, J. D. 1979. Gall formation on *Cirsium arvense* by *Ditylenchus dipsaci*. J. Nematol. 11:16 – 22.

15. Womersley, C. and Smith, L., 1981. Anhydrobiosis in nematodes. I. The role of glycerol, myo-inositol, and trehalose during desiccation. Comp. Biochem. and Physiol. 70B: 579 – 586.

TYLENCHORHYNCHUS

Introduction

These phytonematodes do not appear to be highly specialized in their morphology or behavior. In certain circumstances they may become numerous enough to cause damage. Various species are adapted to particular ecological situations. *T. claytoni* is present in eastern United States, Canada, and Europe in association with turf, potato, maize, tree nurseries, azaleas, and some weeds. *T. cylindricus* is native to arid soils of western United States. *T. annulatus* (= *T. martini*) occurs in tropical and subtropical areas, mostly on rice, sugarcane, and grasses. *T. dubius* is common in Europe on a wide variety of hosts.

Classification

Order Tylenchida; suborder Tylenchina; family Tylenchorhynchidae; subfamily Tylenchorhynchinae.

Morphology

Sexes are separate in most species, and males resemble females except for secondary sex characters. Lengths range from about 0.5 to 1.7 mm with a length/width ratio of 24 – 35. The head is rounded, usually offset, with four to seven annules and with a lightly or strongly sclerotized cephalic framework. Stylet ranges from 16 – 27 μm long and has prominent basal knobs. Esophagus consists of a long, narrow procorpus, prominent median bulb, and a long narrow isthmus. Esophageal glands form a conspicuous pear-shaped basal bulb abutting the intestine. The dorsal esophageal gland orifice is close to the stylet base. Excretory pore is in line with the anterior end of the basal bulb. The vulva is just behind the middle of the body and there are two outstretched gonads. The female tail is blunt, conoid, usually two to three times as long as anal body width. The male tail is pointed, with bursae that extend to the tip (3) (see Fig. 5.1a, p. 76).

Biology

Tylenchorhynchus spp. are common ectoparasites of roots. They feed on epidermal cells in the vicinity of root hairs and in the region of cell elongation. During feeding, an individual remains at one spot for brief periods, less than 10 min. Rapid stylet thrusts persist until a cell wall is breached, then granules move up from esophageal glands to stylet and into the cell. Apparently cell contents are partly digested before the nematode ingests them.

A peculiar behavior of *T. annulatus*, called "swarming," has been described. Nematodes aggregate in masses of wriggling individuals that appear to have sticky cuticles. Electron micrographs of swarming *T. annulatus* show that morphological changes in cuticle have occurred, including swelling of the external cortex, and separation of some layers. In addition, the outermost layer develops projections by which individuals become entangled. Infection with a virus appears to be related to "swarming."

Pathology

Large populations of these nematodes debilitate roots with resulting damage to tops: stunting, chlorosis, and yield reduction.

Interaction with Other Pathogens

Several fungi are associated with *Tylenchorhynchus* spp.: *Phoma medicaginis* var. *pinodella* in peas, *Fusarium roseum* on *Poa pratensis*, and *Aphanomyces euteiches* in peas. Other nematodes apparently suppress reproduction of *Tylenchorhynchus*, possibly because they damage roots sufficiently to reduce available feeding sites. *Pratylenchus, Criconemoides, Belonolaimus, Paratrichodorus, Heterodera*, and *Meloidogyne* are reported to have this effect.

Control

Some species of this genus appear to be more sensitive to nematicides than other nematodes. Weed control is essential for successful use of nematicides. Green manures and the cultivation of various *Compositae* are also useful.

Research

The original genus *Tylenchorhynchus* has been subdivided into several genera. A key to these is published in Ref. (2). Host specificity of different nematodes is illustrated by eggplant—an excellent host for *Meloidogyne incognita* but a poor host for *Tylenchorhynchus brassicae* (1).

References

1. Alam, M. M., Siddiqi, Z. A., Khan, A. M., and Saxena, S. K. 1980. Effect of different cropping sequences on the population of plant parasitic nematodes. Indian J. Nematol. 10:35 – 39.
2. Lewis, S. A. and Golden, A. M. 1981. Description of *Trilineellus clathrocutis* n.g., n.sp. (Tylenchorhynchinae: Tylenchida Thorne, 1949) with a key to species and observations on *Tylenchorhynchus sensu stricto*. J. Nematol. 13:135 – 141.
3. Siddiqi, M. R. 1972. Commonwealth Institute of Helminthology, Descriptions of Plant-parasitic Nematodes, Set 1, No. 7: *Tylenchorhynchus cylindricus*; see also Set 3, No. 39: *T. claytoni*; Set 4, No. 51: *T. dubius*; Set 6, No. 85: *T. annulatus*.

PRATYLENCHUS

Introduction

Nematodes of this genus are destructive pests of many plants, moving freely between roots and soil. A characteristic symptom is the appearance of narrow elongated lesions on root surfaces.

Classification

Order Tylenchida; suborder Tylenchina; superfamily Tylenchoidea; family Pratylenchidae; subfamily Pratylenchinae.

Morphology

In some species sexes are separate; in others males are absent. Nematodes of this genus are elongate, from 340 to 800 μm long, with a length/width ratio of 15 – 35. *Pratylenchus* spp. are recognizable by a flat head, strong cephalic framework, and short thick stylet, 14 – 20 μm long with prominent basal knobs. Esophageal glands overlap the intestine ventrally. An excretory pore opens close to the level of the esophago- intestinal junction. At about 70 – 80% of the body length is the vulva. The single female gonad has a short postvulval sac. Annulations are fine and there are usually four lateral lines, but up to eight are present in some species. A broadly rounded or pointed tail constitutes 3.5 – 9% of the body length. Males, where present, are smaller than females, with bursae that envelop the tail (9) (See Fig. 6.6).

Biology

Pratylenchus spp. are migratory endoparasites, generally with wide host ranges. Eggs, from which second-stage juveniles hatch, are deposited in clusters within roots or in soil. All stages move between soil and roots. Soil populations decline as nematodes invade new roots in late spring and early summer of temperate climates. Nematodes return to soil in late summer and early autumn when roots are declining. Estimates of numbers present must include samples from both plants and soil. Species of this genus prefer coarse textured, sandy soils. At low soil moistures, some species survive well for more than a year in the absence of host crops.

In contrast to many other nematodes, *Pratylenchus* spp. reproduce more actively in roots of plants under stress. For example, populations reach higher levels in alfalfa that is clipped than in alfalfa permitted to grow without clipping. Hosts supplied with minimal nutrient support more nematodes than those with optimal nutrient. The physiological basis for this phenomenon is unknown.

Pathology

Pratylenchus nematodes invade root cortex and kill cells during feeding. Brownish, elongate lesions result, visible at the surface in most, but not

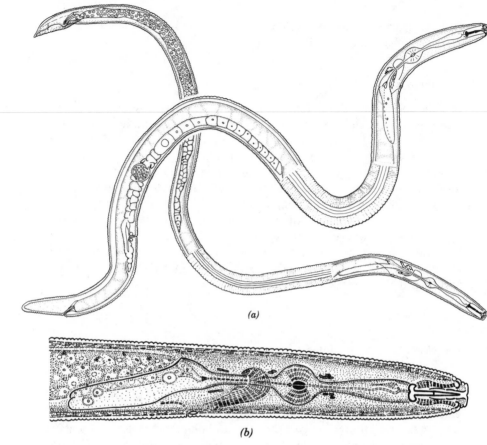

(a)

(b)

Fig. 6.6   *Pratylenchus* spp. (a) Adult male and female of *P. penetrans* ×360; (b) Esophageal region of adult *P. neglectus* ×926.(a) From D. C. M. Corbett. 1973. C.I.H. Descriptions of Plant-parasitic Nematodes, Set 2, No. 25, © Commonwealth Agricultural Bureaux, by permission; (b) From J. L. Townshend, and R. V. Anderson. 1976. *Pratylenchus neglectus* (= *P. minyus*). C.I.H. Descriptions of Plant-parasitic Nematodes, Set 6, No. 82, © Commonwealth Agricultural Bureaux, by permission.

in all hosts. As conditions become unfavorable for the nematodes, they leave and either move to new locations within the root or leave the plant. Some damage within the stele may also result, and an entire root may be killed. Tops of infected plants show symptoms of root damage, such as wilting, yellow leaves, twig dieback, and stunted growth. Peanuts may suffer damage to pegs and shells of pods, causing loss of yield. Severe infestations may kill host plants (see Fig. 6.7).

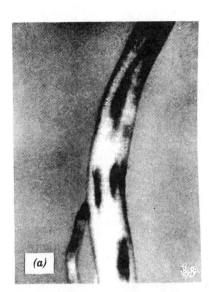

Fig. 6.7 Pathology induced by *Pratylenchus* spp. (a) Lesions on adventitious root of strawberry, with a small feeder root girdled at base following attack by *P. penetrans*. (b) Longitudinal section through cortex of a feeder root of *Citrus sunki*, 8 days after being inoculated with 100 *P. coffeae*. N, nematode) ×110. (a) (From D. C. M. Corbett. 1973. *Pratylenchus penetrans*. C.I.H. Descriptions of Plant-parasitic Nematodes, Set 2, No. 25, © Commonwealth Agricultural Bureaux, by permission.); (b) (From C. S. Huang, and Y. C. Chiang. 1976. Pathogenicity of *Pratylenchus coffeae* on Sunki Orange. Plant Dis. Rep. 60:957 – 960, by permission.)

Interaction with Other Pathogens

Lesions in the cortex are excellent sites for other pathogens. In wilt of peppermint and potato, *Pratylenchus* and *Verticillium* each enhances the effect of the other. Split-root experiments in peppermint showed that the fungus benefits from the presence of *Pratylenchus* elsewhere in the root. Other fungi found in roots containing these nematodes include *Phytophthora*, *Pythium*, *Peronospora*, *Aphanomyces*, *Cylindrocarpon*, *Fusarium*, and *Trichoderma*. In most cases, the nematode enhances fungal growth and consequent damage to the plant. But antagonistic relations have also been found in which each pathogen is inhibited by the other (2, 4). In addition to interactions with fungi, some bacterial infections are enhanced by *Pratylenchus*, including bacterial wilt in tomato caused by *Pseudomonas* and hairy root of roses (*Agrobacterium*).

Control

Nematicides are widely used to control populations of *Pratylenchus* spp. where the economics of crop production is favorable. Hot water treatment of fruit tree transplants and chemical dips of bulbs have also been used. Crop rotation is effective, but care must be taken to determine whether the nematodes present can reproduce on the alternate crop. In addition, weed control is essential. Fallowing also may be effective.

Tagetes spp. (marigolds) are antagonistic to *Pratylenchus*. Extensive tests around the world have shown that control by *Tagetes* is feasible in certain circumstances. A major problem is that no market exists for marigolds. Plant breeding against this nematode offers promise for control. Work is currently in progress on alfalfa, peaches, rice, citrus, and other crops. Resistant and tolerant varieties of corn, millet and alfalfa have been produced.

Research

Research on *Pratylenchus* is concentrated in several areas: relation between numbers of invading nematodes and damage, effect of plant age on tolerance, genetic variability of host, varietal sensitivity to attack, and most prominently, interactions of several pathogens in the same plant. Both intensity of attack and age of plant influence disease severity. Low numbers of *P. penetrans* increase shoot and root growth of navy bean but higher numbers decrease them. Some varieties are severely damaged but others are tolerant (3). Nematode attack on seedlings is more damaging than attack on older plants. Nematode reproduction on rice is inhibited in inoculations of 32-day-old or older plants in comparison with inoculations on seedlings (7).

A troublesome disease of potatoes, "potato early dying," results from interaction of *Verticillium dahliae* and *Pratylenchus* spp. Field experiments in Ohio showed that *P. crenatus* did not interact with the fungus to produce disease, but *P. penetrans* and *P. scribneri* did (8).

A significant development in the study of phytonematodes has been the refinement of techniques to cultivate nematodes under aseptic conditions. Cultures of roots, callus, and carrot discs have proved useful. A recent example of the use of carrot disc is described in Ref. (1) and an example of the use of corn root cultures appears in Ref. (6).

In addition to interactions between phytonematodes and fungal pathogens, there is an interaction between nematodes and mycorrhizae. *P. penetrans* inhibits root colonization of *Phaseolus vulgaris* by *Glomus fasciculatum* and the nematodes increase more rapidly on mycorrhizal roots. The fungus, however, reproduces poorly in the presence of the nematode (4). Many field infections consist of more than one phytonematode species; these may interact with each other. *P. coffeae* and *Tylenchulus semipenetrans* do not occur together in Florida orchards; experimental inoculations indicate that the two nematodes are mutually exclusive (5). Combinations of cyst nematodes in microplots containing *P. penetrans* are not stable. Lesion nematodes do not maintain their populations in combination with certain species of *Globodera* and *Heterodera*. But not every species of cyst nematode is effective in suppressing lesion nematodes (2). This finding probably reflects a complex interaction among host plants, predators and parasites of nematodes, and species of phytonematodes.

References

1. Chitambar, J. J. and Raski, D. J. 1985. Life history of *Pratylenchus vulnus* on carrot discs. J. Nematol. 17:235 – 236.

2. Decker, H. & Seidel, D., 1981. [Observations on the displacement of migratory root nematodes of the genus *Pratylenchus* by cyst nematodes (*Globodera, Heterodera*).] Arch. Phytopathol. Pflanzenschutz 17:39 – 46 (German, English).

3. Elliott, A. P. and Bird, G. W. 1985. Pathogenicity of *Pratylenchus penetrans* to navy bean (*Phaseolus vulgaris* L.). J. Nematol. 17:81 – 85.

4. Elliott, A. P., Bird, G. W., and Safir, G. R. 1984. Joint influence of *Pratylenchus penetrans* (Nematoda) and *Glomus fasciculatum* (Phycomyceta) on the ontogeny of *Phaseolus vulgaris*. Nematropica 14:111–119.

5. Kaplan, D. T. and Timmer, L. W. 1982. Effects of *Pratylenchus coffeae* — *Tylenchulus semipenetrans* interactions on nematode population dynamics in citrus. J. Nematol. 14: 368 – 373.

6. Meyer, A. J. 1985. Mass culture of *Pratylenchus zeae* (Nematoda: Pratylenchinae) on excised maize roots growing on sterile nutrient agar. Phytophylactica 16:259 – 261.

7. Prasad, J. S. and Rao, Y. S. 1983. Influence of age of rice plant at the time of inoculation of *Pratylenchus indicus* Das on the multiplication of the nematode and yields of host plant. Acta Oecol. 4:309 – 312.

8. Riedel, R. M., Rowe, R. C., and Martin, M. J. 1985. Differential interaction of *Pratylenchus crenatus, P. penetrans*, and *P. scribneri* with *Verticillium dahliae* in potato early dying disease. Phytopathology 75:419 – 428.

9. Siddiqi, M. R. 1972. C.I.H. Descriptions of Plant-parasitic Nematodes Set 1, No. 6: *Pratylenchus coffeae*; see also Set 2, No. 25: *P. penetrans*; Set 3, No. 37: *P. vulnus*; Set 6, No. 77: *P. zeae*; Set 6, No. 82 *P. neglectus (= P. minyus)*; Set 6, No. 89: *P. brachyurus*.

RADOPHOLUS

Introduction

A species of this genus, *R. similis*, is a destructive pest of banana, citrus, and other crops, especially in warm climates. During the 1950's, these nematodes threatened Florida citrus production. Interest in saving the citrus industry stimulated general awareness of phytonematodes in agriculture.

Classification

Order Tylenchida; suborder Tylenchina; superfamily Tylenchoidea; family Pratylenchidae; subfamily Radopholinae.

Morphology

Sexes are separate. Females of *R. similis* average 690 μm long with a ratio of length to width of about 27. The slightly offset, hemispherical head has a well-developed cephalic framework. Stylet is strong, about 19 μm long, with prominent rounded knobs. Esophageal glands (three separate lobes) overlap the intestine dorsally. Excretory pore is at the level of the esophago-intestinal valve. An anterior and a posterior gonad join the vulva just behind the body midpoint. Well-marked annulations with four lateral lines are present. The tail forms about 10% of the body length. Males average 585 μm long and are degenerate, with imperfectly developed esophagus and stylet. The male head is rounded and markedly offset in contrast to the slightly offset female head. There is a single testis and bursae envelop about two-thirds of the tail (6) (see Fig. 6.8).

Biology

R. similis is called the "burrowing nematode" in reference to its active movement and cell destruction within roots. Both juveniles and adult females invade roots. Adult females remain within roots until tissues are badly damaged. Sexual reproduction is generally required, but parthenogenesis also occurs. A number of races are known that fall into two major subdivisions. One group attacks citrus and banana and the other attacks banana but not citrus. Another species of *Radopholus* damages rice. This nematode is hard to control on citrus because it follows roots deep into soil, down to 3.7 m.

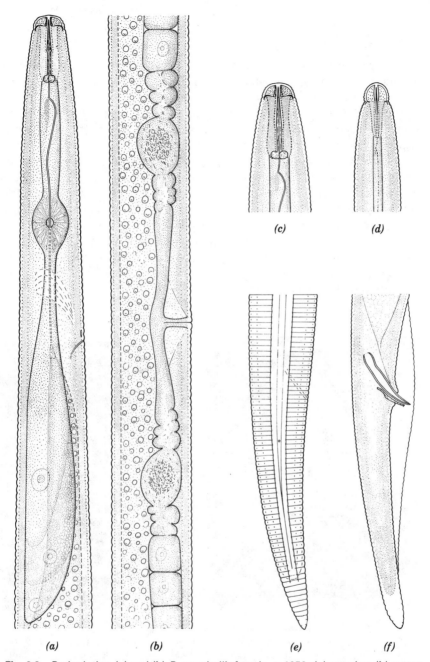

(a) (b) (c) (d) (e) (f)

Fig. 6.8 *Radopholus.* (a) and (b) *R. neosimilis* female, ×1050; (a) anterior; (b) center of body. (c) – (f) *R. similis*, ×985; (c) and (e) female; (d) and (f) male; (from S. A. Sher. 1968. Revision of the genus *Radopholus*, Thorne 1949 (Nematoda: Tylenchoidea). Proc. Helminthol. Soc. Wash. 35:219 – 237, by permission of Helminthological Society of Washington).

Pathology

R. similis is called the burrowing nematode in reference to its behavior within roots. In banana it enters cortical parenchyma where it moves about actively, destroying cells as it feeds. Cavities develop and enlarge, but do not cross the endodermis. Reddish brown lesions appear throughout the cortex. Anchor roots of banana are destroyed and plants topple readily, especially when heavy fruit is present.

R. similis causes "spreading decline" of citrus. Infection moves rapidly through an orchard. The nematode tunnels through root cortex and invades cambial and phloem layers of the stele. Pericycle cells are stimulated to divide, wound gum is deposited in the cortex, and a small gall develops, the surface of which cracks. More than 700 individuals have been recovered from a single gall. A heavily infected citrus tree becomes stunted and unthrifty and may eventually die. This pathogen induces similar effects in black pepper and avocado. Figure 6.9 shows typical lesions induced by this nematode on banana and citrus.

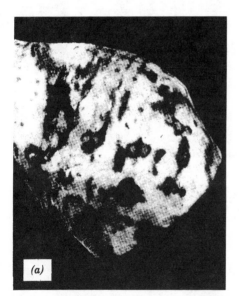 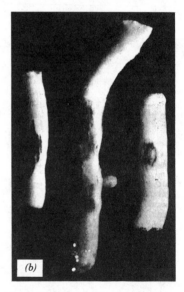

Fig. 6.9 Root lesions (*R. similis*). (a) Pared banana rhizome showing lesions surrounding embedded roots. (b) Spreading decline—citrus feeder roots. (a) (From K. J. Orton Williams and M. R. Siddiqi. 1973. *Radopholus similis*. C.I.H. Descriptions of Plant-parasitic Nematodes, Set 2, No. 27. © Commonwealth Agricultural Bureaux, by permission.) (b) (From J. M. Webster, ed. 1972. Economic Nematology. Academic Press, New York, by permission.)

Table 6.1 Symptoms of Tree Damage to Citrus by *Tylenchulus semipenetrans* and *Radopholus similis* in Florida[a]

Slow Decline: *Tylenchulus semipenetrans*	Spreading Decline: *Radopholus similis*
Foliage sparse, often dull gray-green or bronze-green. Leaves smaller than normal, upright and cupped. Fruit set reduced, fruit smaller than normal. New flushes of growth are reduced. Trees often wilt during midday when healthy trees do not	Thin, chlorotic foliage, small leaves, twig dieback, absence of older leaves or of heavy bloom in spring. Extremely poor fruit set. Less foliage than in slow decline
Diseased trees not separated by sharp boundary from healthy. Feeder roots have an encrusted appearance from sand embedded in egg sacs	Diseased trees in a clump. Affected area enlarges each year. Well-defined margin between diseased and healthy trees. Feeder roots not encrusted

[a] Adapted from Hannon (1).

Symptoms of "spreading decline" are different from "slow decline" of citrus resulting from infection with *Tylenchulus semipenetrans*. Table 6.1 compares symptoms in the two diseases.

Interaction with Other Pathogens

Extensive necrotic areas in citrus roots induced by *Radopholus* are readily occupied by microorganisms, including the pathogenic fungi *Fusarium* and *Sclerotium*. *Fusarium oxysporum f. cubense* causes an important disease of banana, Panama wilt. This fungus does not invade uninjured roots, but in the presence of *R. similis*, it reaches the stele and the root atrophies beyond the point of colonization.

Control

Infected banana sets can be prepared for planting by cutting away discolored lesions. This practice, together with nematicide dips, severely limits nematode distribution during replanting. Hot water treatment of banana sets is also effective. Spreading decline of citrus in Florida during the 1950s and 1960s generated much excitement among growers and others. Trees showing damage were removed from infected areas and from a perimeter of undamaged rows; the soil was treated with heavy doses of nematicides before replanting. The disease was gradually controlled. Genetic resistance now offers promise for control of *Radopholus* in citrus, tea, and banana.

Research

Investigators have had to contend with populations of indistinguishable morphology but different host ranges. Nematologists have attempted recently to characterize such populations by several methods. A group of *Radopholus* populations that attack citrus all have a chromosome complement of $n = 5$, whereas those that attack banana have $n = 4$. This difference suggests that two species rather than races are involved (3). The authors also found that two populations differing in host preferences cannot be crossed. Therefore they conclude that these populations belong to sibling species; that is, genetically isolated but morphologically identical species (4). Earlier work demonstrated differences in enzymes between the citrus and banana races (2).

"Spreading decline" of citrus (*R. similis*) in Florida was confined to well-drained sands whereas "slow decline" (*Tylenchulus semipenetrans*) was not restricted to this soil type. Experimental tests showed that *R. similis* prefers deep, well-drained sands, whereas *T. semipenetrans* is indifferent to soil type (5).

References

1. Hannon, C. I. 1962. The occurrence of the citrus-root nematode, *Tylenchulus semipenetrans* Cobb in Florida. Plant Dis. Rep. 46:451 – 455.

2. Huettel, R. N., Dickson, D. W., and Kaplan, D. T. 1983. Biochemical identification of two races of *Radopholus similis* by starch gel electrophoresis. J. Nematol. 15:338 – 344.

3. Huettel, R. N., Dickson, D. W., and Kaplan, D. T. 1984. Chromosome numbers of populations of *Radopholus similis* from North, Central, and South America, Hawaii and Indonesia. Rev. Nématol. 7:113 – 116

4. Huettel, R. N., Dickson, D. W., and Kaplan, D. T. 1984. *Radopholus citrophilus* sp. n. (Nematoda), a sibling species of *Radopholus similis*. Proc. Helminthol. Soc. Wash. 51: 32 – 35.

5. O'Bannon, J. H. and Esser, R. P. 1985. Citrus Declines caused by Nematodes in Florida. I. Soil Factors. Nematol. Circ. No. 114. Div. Plant Ind., Florida Dept. of Agriculture and Consumer Service, Gainesville, 4 pp.

6. Orton Williams, K. J. and Siddiqi, M. R. 1973. Commonwealth Institute of Helminthology, Descriptions of Plant- parasitic Nematodes, Set 2, No. 27: *Radopholus similis*.

HIRSCHMANNIELLA

Introduction

Hirschmanniella oryzae, formerly named *Radopholus oryzae*, is an important parasite of rice.

Classification

Order Tylenchida; suborder Tylenchina; superfamily Tylenchoidea; family Pratylenchidae; subfamily Pratylenchinae.

Morphology

Long, narrow nematodes, measuring 0.9 – 4.2 mm in length, with a length/width ratio of 40 – 60. Sexes are separate, males resemble females except for secondary sex characters. Lip region is low, flattened, with rounded edges in some species and hemispherical in others, and not offset from body. Head skeleton is strongly developed, with outer margins extending posteriorly several body annules. A strong stylet is up to 40 μm long. Esophageal glands overlap the intestine ventrally for some distance. Excretory pore is at level of esophago-intestinal junction or somewhat anterior to it. An anterior and a posterior gonad join at the midbody vulva. The rounded or conoid tail may have a spike (= mucro). Bursae envelop only part of male tail. There are four incisures in the lateral field. Nematodes of this genus have a unique structure described as "large elongated thin-walled cells of unequal size arranged in tandem between the body muscle and the intestinal tract" (5, 6). See Fig. 6.10 for a diagram of *H. oryzae*.

Biology

These are migratory internal parasites of rice and other plants of aquatic or moist situations in both tropical and temperate regions. Weeds in paddy fields carry the population over between rice crops.

Pathology

Hirschmanniella spp. nematodes kill cells of the cortex and dissolve cell walls, thereby causing large cavities to form. Rice plants suffer reduction of growth and yield. Under some conditions new roots may permit partial recovery.

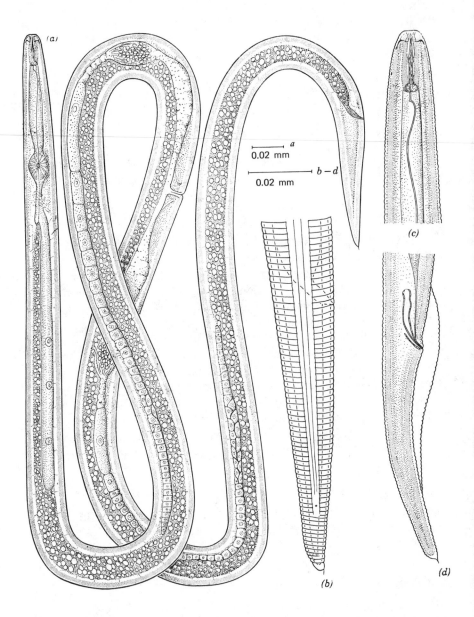

Fig. 6.10 *Hirschmanniella oryzae.* (a) Female ×430; (b) Female posterior region; (c) Male, anterior end; (d) Male, posterior. (b-d) ×850. (From S. A. Sher. 1968. Revision of the genus *Hirschmanniella* Luc & Goodey, 1963 (Nematoda:Tylenchoidea). Nematologica 14:243 – 275, by permission of E.J. Brill.)

Interaction with Other Pathogens

None reported.

Control

Nematicides have been employed successfully in the field to reduce nematode populations before transplanting rice seedlings.

Research

Much attention has been given to this organism on rice in India and Japan. A general review of research on rice root nematodes of India is presented in Ref. (2). A recent key to species of the world is contained in Ref. (1). A search for genetic resistance is reported in Ref. (3).

Nitrogen fixation by microorganisms other than nodulating bacteria is important in rice cultivation. Rice root nematodes interfere with nitrogen fixation, presumably by depressing root exudations upon which nitrogen-fixing bacteria feed (4).

References

1. Ebsary, B. and Anderson, R. V. 1982. Two new species of *Hirschmanniella* Luc & Goodey, 1963 (Nematoda: Pratylenchidae) with a key to the nominal species. Can. J. Zool. 60: 530 – 535.
2. Edward, J. C., Sharma, N. N., and Agnihothrudu, V. 1985. Rice root nematode (*Hirschmanniella* spp.) — a review of the work done in India. Current Sci. 54:179 – 182.
3. Ramakrishnan, S., Varadharajan, G. & Sutharsan, P. D. 1984. TKM9 is resistant to rice-root nematode. Int. Rice Res. News. 9:20.
4. Rinaudo, G. & Germani, G. 1981. Effect of the nematodes *Hirschmanniella oryzae* and *H. spinicaudata* on the N_2 fixation in the rice rhizosphere. Rev. de Nématol. 4:171 – 172.
5. Sher, S. A. 1968. Revision of the genus *Hirschmanniella* Luc & Goodey, 1963 (Nematoda:Tylenchoidea). Nematologica 14:243 – 275.
6. Siddiqi, M.R., 1973. Commonwealth Institute of Helminthology Descriptions of Plant-parasitic Nematodes, Set 2, No. 26: *Hirschmanniella oryzae*; see also Set 5, No. 68: *H. spinicaudata*.

NACOBBUS

Introduction

This genus occurs in western N. and S. America, causing galls on roots of many agricultural and weed hosts.

Classification

Order Tylenchidae; suborder Tylenchina; superfamily Tylenchoidea; family Pratylenchidae; subfamily Nacobbinae.

Morphology

Sexes of these nematodes are markedly different. Mature females are swollen, spindle-shaped, with elongated neck and tail and a total body length of 1 mm or longer. Males are elongate, about 0.9 mm long, with small caudal alae. Length/width ratio is close to 30. Immature males and females are elongated and motile. In all stages the head is not set off. Cephalic framework and stylet are well developed. The median bulb in adults has prominent crescentic plates whereas juveniles have normal plates. Esophageal glands are elongate, overlapping the intestine dorsally. Ovary is single, and the mature female is packed with embryonated eggs. A taxonomic review of the genus is published (4) and diagrams of life stages are presented in Fig. 6.11.

Biology

The life cycle of *Nacobbus* differs from that of other sedentary endoparasites. Infective second-stage juveniles enter roots and may or may not induce swelling. They move intracellularly, causing extensive necrosis. Third- and fourth-stage females are motile, migrate from roots to soil, and reenter roots where they induce galls. Adult females stimulate formation of a spindle-shaped syncytium upon which they feed. Adult females swell, but retain an elongate tail that reaches to the root surface as the nematodes mature. Eggs are deposited into a gelatinous egg sac. The elongate male leaves the root to move about in search of a female. Several males may be found within a single egg sac.

Pathology

Top symptoms resemble those in plants infected with other root pathogens: reduced growth, diminished ability to withstand water stress, and decreased yield. Extensive necrosis resulting from intracellular migration of juveniles severely damages roots. Adult females induce partial cell-wall dissolution connecting hundreds of cells together into a syncytium at one end of which the nematode remains, feeding on cell contents. Cells of the syncytium are slightly enlarged and thin-walled, and display signs of intense metabolism. Wall ingrowths associated with giant cells of *Meloidogyne* and syncytia of

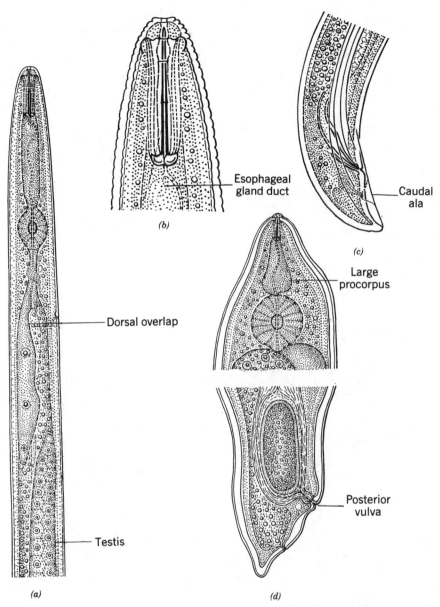

Fig. 6.11 *Nacobbus.* (a) Male, anterior end ×565; (b) male, head region ×1740; (c) male, posterior; [×960]; (d) adult female, anterior and posterior ends [×300]; (e), (f) stages of maturing females [×49]; (g) mature female with eggs [×35]; (h) dissected gall [×17]. (a) – (d) From G. Thorne 1961. Principles of Nematology, 553 pp., by permission of McGraw-Hill Book Co.; (e) – (g) From S. A. Sher. 1970. Revision of the Genus *Nacobbus* Thorne and Allen, 1944 (Nematoda: Tylenchoidea). J. Nematol. 2:228 – 235, by permission;) (h) from J. B. Goodey. 1963. Soil and Freshwater Nematodes, by permission of John Wiley & Sons.

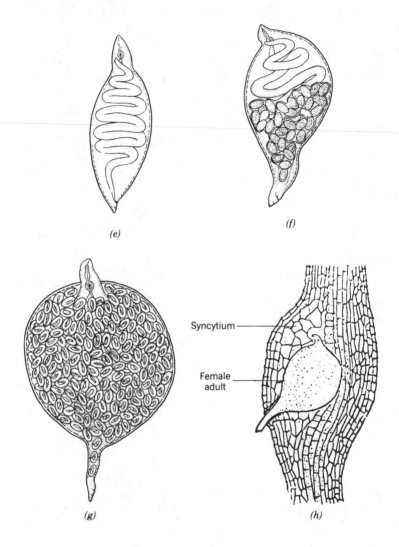

(e)

(f)

Syncytium ———

Female
adult ——

(g)   (h)

Fig. 6.11 (Continued)

Heterodera are not present in *Nacobbus* infections. A syncytium may extend up to 3 mm long by 2 mm wide. Large amounts of starch accumulate
in these cells and in the surrounding gall tissue. The dark color of iodine-
stained starch in galls induced by *Nacobbus* readily distinguishes them from
galls induced by *Meloidogyne*.

Interaction with Other Pathogens

There are no reports of this although microorganisms invade necrotic regions and contribute to decay.

Control

Standard practices of crop rotation and use of nematicides are employed. Until recently there has been little interest in finding genetic resistance to this organism. Recent reports of extensive infestation of potatoes in South America should stimulate work in plant breeding.

Research

See Ref. (2) for a review of research progress on this organism. The life history of *Nacobbus* on tomato is well illustrated in Ref. (1). A careful study of the syncytium, with special attention to the occurrence of numerous plasmodesmata between cells appears in Ref. (3).

References

1. Clark, S. A. 1967. The development and life history of the false root-knot nematode, *Nacobbus serendipiticus*. Nematologica 13:91 – 101.
2. Jatala, P. 1978. Review of the false root-knot nematode (*Nacobbus* spp.). Research Progress, International Potato Center. Report of the 2nd Planning Conference on the Developments in the Control of Nematode Pests of Potatoes, Lima, Peru, 13 – 17 November, 1978.
3. Jones, M. G. K. and Payne, J. L. 1977. The structure of syncytia induced by the phytoparasitic nematode *Nacobbus aberrans* in tomato roots, and the possible role of plasmodesmata in their nutrition. J. Cell Sci. 23:299 – 313.
4. Sher, S. A. 1970. Revision of the Genus *Nacobbus* Thorne and Allen, 1944 (Nematoda: Tylenchoidea). J. Nematol. 2:228 – 235.

HOPLOLAIMUS

Introduction

This genus is one of a group equipped with robust spears and strong cephalic frameworks with which these nematodes penetrate plant cells. Species of *Hoplolaimus* have a wide host range, including grasses, cereals, soybeans, corn, cotton, sugarcane, and trees.

Classification

Order Tylenchida; suborder Tylenchina; superfamily Tylenchoidea; family Hoplolaimidae; subfamily Hoplolaiminae.

Morphology

Sexes are separate in most species; males resemble females except for secondary sex characters. *Hoplolaimus* spp. are cylindrical, somewhat plump nematodes ranging in length from about 1 to 2 mm with a length/width ratio of 22 – 38. Lip region is set off and some or all of its annulations are subdivided by longitudinal striations. Cephalic framework is thick, yellowish in some species. Stylet is massive, ranging from 33 to 52 μm long, with large knobs marked by anterior projections. Esophageal glands overlap the intestine dorsally and laterally. Distance between dorsal gland orifice and stylet is less than one-fourth of stylet length. Location of excretory pore varies among *Hoplolaimus* spp., ranging from the level of the midmedian bulb to a level behind the esophago-intestinal junction. The vulva is just behind the midpoint of the body and there are two gonads. Phasmids in this genus are large, not opposite one another. One may be in the anterior third of the body and the other in the posterior region. Cuticle annulations are well marked, with four or fewer lateral incisures. The female tail is short and round, with annulations extending to the terminus. Male tails are pointed, with bursae reaching to the tip (2). A typical example of this genus is presented in Fig. 6.12.

Biology

These are mostly ectoparasites but some penetrate roots completely.

Pathology

Cortex cells are destroyed during feeding, and necrotic lesions may be extensive, including some cells at a distance from the nematode. Individuals inside roots may cross the endodermis to reach the stele where phloem and xylem tissues may be damaged.

Interaction with Other Pathogens

Fusarium moniliforme, associated with *H. indicus* on maize, causes wilting when the nematode is present, but not when it is absent. *H. seinhorsti* has been associated with *F. oxysporum* in cotton, and *H. galeatus* with the

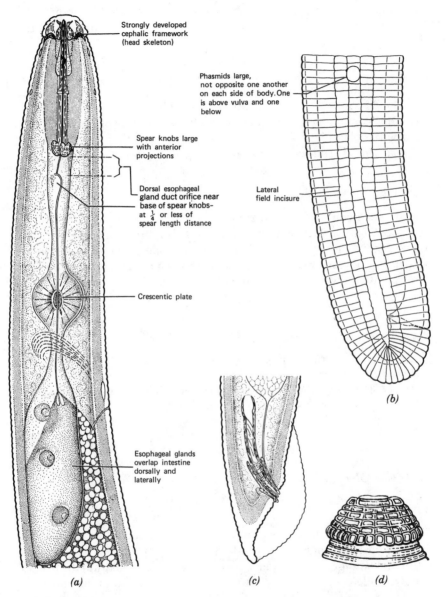

Strongly developed cephalic framework (head skeleton)

Phasmids large, not opposite one another on each side of body. One is above vulva and one below

Spear knobs large with anterior projections

Dorsal esophageal gland duct orifice near base of spear knobs— at $\frac{1}{4}$ or less of spear length distance

Lateral field incisure

Crescentic plate

Esophageal glands overlap intestine dorsally and laterally

(a)   (c)   (d)

(b)

Fig. 6.12 *Hoplolaimus*. (a) *H. stephanus*, female, anterior; (b) female, posterior; (c) male, posterior (a – c ×800). (From S. A. Sher. 1963. Revision of the Hoplolaiminae (Nematoda) II. *Hoplolaimus* Daday, 1905 and *Aorolaimus* n.gen. Nematologica 9: 267 – 295, by permission of E. J. Brill); (d) lip region of *H. coronatus* (×1640). (From G. Thorne. 1961. Principles of Nematology. McGraw-Hill, by permission.)

same fungus in peach. Lesions induced by these nematodes are probably excellent portals of entry and good substrates for many pathogenic organisms.

Control

Nematicides are effective in controlling populations of *Hoplolaimus* on turf. Dips containing nematicides are useful to sanitize sprigs of Bermuda grass before planting. Fallowing the soil for three months has been effective in India. But *H. columbus* survived storage in dry soil for 5 years!

Research

Five soybean cultivars with tolerance for *Heterodera glycines* and *Hoplolaimus columbus* were found in a test (1). This represents a new direction in plant breeding. Breeders traditionally seek plants that grow well in the presence of phytonematodes and also inhibit nematode reproduction. Tolerant plants, however, do not suffer damage but permit the parasites to reproduce.

References

1. Boerma, H. R. and Hussey, R. S. 1984. Tolerance to *Heterodera glycines* in soybean. J. Nematol. 16:189 – 196.
2. Orton Williams, K. J. 1973. Commonwealth Institute of Helminthology, Descriptions of Plant-parasitic Nematodes, Set 2, No. 24: *Hoplolaimus galeatus*; see also Set 3, No. 33: *H. pararobustus*; Set 5, No.66: *H. indicus*; Set 6, No. 76: *H. seinhorsti*; and Set 6, No. 81: *H. columbus*.

SCUTELLONEMA

Introduction

This genus belongs to a group called "spiral nematodes" in reference to their coiled appearance after gentle heat treatment and in soil extracts. *Scutellonema* spp. are important agricultural pests, especially in the tropics. Hosts include coconut, yams, cotton, banana, corn, and ornamentals.

Classification

Order Tylenchida; suborder Tylenchina; superfamily Tylenchoidea; family Hoplolaimidae; subfamily Hoplolaiminae.

Morphology

In some species males are rare or absent; in others they are abundant and resemble females in general appearance. These nematodes are close to 1 mm long, with a length/width ratio of 27 – 32. They assume a coiled shape when killed by heat. Head is offset, hemispherical or conoid, and cephalic skeleton is well developed. The stylet is strong, about 26 – 30 μm long with prominent knobs. Cuticle annulations are well marked in both sexes, with distinct lateral lines. Prominent phasmids, 3 – 4 μm across and situated at or just anterior to the anus, are diagnostic of this genus. Esophageal glands overlap the intestine for a short distance dorsally and dorsolaterally. The vulva is just posterior to the midbody. One gonad extends anteriorly and the other posteriorly. The tail is broadly rounded with annulations encircling it entirely. Males, where present, have bursae reaching to the tail tip (8). Figure 6.13 illustrates the diagnostic characters.

Biology

Scutellonema spp. are mainly ectoparasites but may also be found entirely within roots. The life cycle of *S. cavenessi* in the Sahel of Senegal is adapted to alternating seasons of rainfall and drought. At the end of the rainy season the soil contains fourth-stage juveniles and adults. These coil and become quiescent. They survive the dry season until rains begin again, when juveniles molt to become adults. The nematodes mate, and females deposit eggs. Second-stage juveniles hatch, invade roots, and continue to develop. The nematodes return to soil as fourth-stage juveniles (3).

Pathology

Scutellonema destroys cells of the cortex during feeding. It may be an ectoparasite or invade roots completely. Some plants form periderm tissue with suberized cell walls around the injury. This genus is most important in the tropics where it damages stored yams (*Dioscorea*). Infected tubers develop a dry rot in the absence of bacterial infection or a wet rot when bacteria invade the wounds. Increased weight loss during storage as well as loss of tuber tissue cause severe damage to the stored crop (1). Cortical lesions in roots of lily and other hosts are often extensive.

Interaction with Other Pathogens

No specific interactions of *Scutellonema* and bacteria or fungi have been reported. Damage to legume roots diminishes nodulation by *Rhizobium*; this has been found in plants infected with *Scutellonema*.

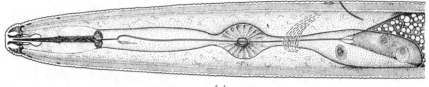

(a)

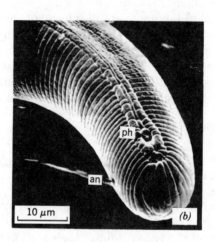

(b)

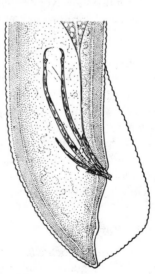

(c)

Control

Hot water treatment and soaking in nematicides control *Scutellonema* on yams. Use of nematicides in the field is also effective.

Research

Cells directly penetrated by the stylet have ultrastructural changes and cells at some distance show darkening but without ultrastructural changes. A membranous network surrounds the stylet within a cell. Macrotubules appear, and the cytoplasm shows characteristics of senescence as illustrated in Fig. 6.14. Cells eventually die after penetration (7).

Scutellonema cavenessi is a problem in the Sahel of Senegal, where it damages peanut and soybean. Infected plants have reduced dry weights of tops and roots, and reduced nitrogen fixation. These effects are greater on soybean than on peanut (4). Just enough rain falls in 3 months to support a crop. For 9 months thereafter the soil dries. Almost all of the nematodes remain in the upper 20 cm of the dry soil in a coiled condition. During gradual desiccation from 100% relative humidity to that of dry soil (97.7% R.H.), over 80% of the nematodes are coiled. Most of them survive exposure to 0% humidity for 1 month. Obviously, this nematode and probably others in the same area are well adapted to survive soil desiccation (2).

Experiments on the interaction of *Scutellonema cavenessi* and the mycorrhizal fungus, *Glomus mossae*, had surprising results. As expected, the fungus improved growth of soybeans whereas the nematode depressed it. But in joint infections, although nematode reproduction was enhanced, damage was reduced. As Germani et al conclude, these interrelations are complex and probably differ from one association to the next (5). Research on cotton offers another illustration of the complexity of interrelations among phytonematodes. The authors inoculated cotton seedlings with all possible combinations of *Hoplolaimus columbus*, *Scutellonema brachyurum*, and *Meloidogyne incognita*. Combination of *Hoplolaimus* and *Scu-*

Fig. 6.13 *Scutellonema.* (a) *S. validum*, female anterior (×660). (Note esophageal glands overlapping intestine dorsally and laterally); (b) *S. cavenessi*, posterior of female (×595), arrow denotes large phasmid; these are opposite in anal region. (c) *S. validum*, male anterior and posterior (×1000). (a) and (c) from S. A. Sher. 1963. Revision of the Hoplolaiminae (Nematoda) III. *Scutellonema* Andrassy, 1958. Nematologica 9:421–443, by permission of E.J. Brill; (b) from A. T. De Grisse. 1977. De Ultrastruktuur van het Zenuwstelsel in de Kop van 22 Soorten Plantenparasitaire Nematoden, behorende tot 19 Genera (Nematoda: Tylenchida). Rijksuniversiteit Gent, Belgium, 420 pp., Fakulteit van de Landbouwwetenschappen Laboratorium voor Dierkunde, by permission.

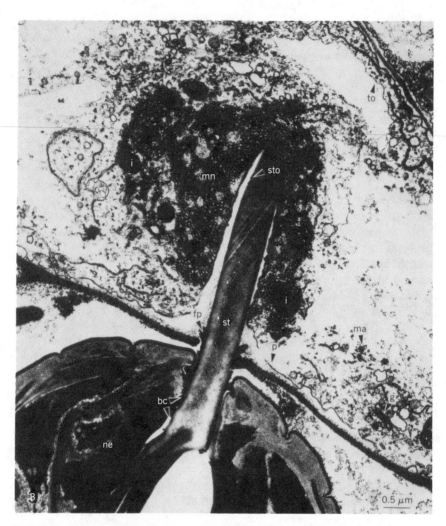

Fig. 6.14 Electron micrograph (×23,000) of cross section through penetrated root tissue with stylet of *Scutellonema brachyurum* inserted into a cortical cell. A feeding plug (fp) forms where the nematode's (ne) stylet (st) penetrates the cell wall. The material forming the feeding plug seems to emanate from the buccal cavity (bc). Associated with the stylet is a membraneous network (mn) containing spherical inclusion bodies (i) that have the same ultrastructure and electron density as material found in the stylet's orifice (sto). Macrotubules (ma) occur through the penetrated cell. Tonoplast (to) and plasmalemma (p) are discontinuous and Golgi bodies (g) have irregular membranes. (From A. C. Schuerger and M. A. McClure. 1983. Ultrastructural changes induced by *Scutellonema brachyurum* in potato roots. Phytopathology 73: 70 – 81, by permission of American Phytopathological Society.)

tellonema stimulated the reproduction of both nematodes, but *Meloidogyne* suppressed *Scutellonema* (6).

References

1. Adesiyan, S. O., Odihirin, R. A., & Adeniji, M. O. 1975. Economic losses caused by the yam nematode, *Scutellonema bradys*, in Nigeria. U.S.D.A. Plant Dis. Rep. 59:477 – 480.

2. Demeure, Y. 1980. Biology of the plant-parasitic nematode *Scutellonema cavenessi* Sher, 1964: anhydrobiosis. Rev. Nématol. 3:283 – 289.

3. Demeure, Y., Netscher, C., and *Quénéhervé*. 1980. Biology of the plant-parasitic nematode *Scutellonema cavenessi* Sher, 1964: reproduction, development and life cycle. Rev. Nématol. 3:213 – 225.

4. Germani, G. 1981. Pathogenicity of the nematode *Scutellonema cavenessi* on peanut and soybean. Rev. Nématol. 4:203 – 208.

5. Germani, G., Ollivier, B., & Diem, H.G. 1981. Interaction of *Scutellonema cavenessi* and *Glomus mosseae* on growth and N_2 fixation of soybean. Rev. Nématol. 4:277 – 280.

6. Kraus-Schmidt, H. and Lewis, S.A. 1981. Dynamics of concomitant populations of *Hoplolaimus columbus*, *Scutellonema brachyurum*, and *Meloidogyne incognita* on cotton. J. Nematol. 13:41 – 48.

7. Schuerger, A. C. and McClure, M. A. 1983. Ultrastructural changes induced by *Scutellonema brachyurum* in potato roots. Phytopathology 73:70 – 81.

8. Siddiqi, M. R. 1972. Commonwealth Institute of Helminthology, Descriptions of Plant-parasitic Nematodes, Set 1 No. 10: *Scutellonema bradys*; see also Set 4, No 54: *S. brachyurum*.

ROTYLENCHUS

Introduction

Nematodes of this genus, together with *Scutellonema* and *Helicotylenchus*, are called "spiral nematodes" because they coil into a C-shape or spiral when killed by heat. Hosts of *Rotylenchus* spp. include ornamentals, vegetables, conifers, and grasses. Species occur in Europe, Soviet Union, North America, and Egypt; the genus is probably cosmopolitan.

Classification

Order Tylenchida; suborder Tylenchina; superfamily Tylenchoidea; family Hoplolaimidae; subfamily Rotylenchinae.

Morphology

Some species of *Rotylenchus* have separate sexes, whereas in others males are unknown. Females are 1 – 2 mm long, with a length/width ratio of

32 – 40. The female head is hemispherical and may be slightly offset or not. Lips of males are more distinctly offset and elevated than in females. One or more labial annules are subdivided by longitudinal striations, giving a tiled appearance. Cephalic framework is strong and the stylet is robust, measuring 33 – 50 μm long. Esophageal glands overlap the intestine dorsally for a short distance. The excretory pore is at the level of the esophago-intestinal junction. The orifice of the esophageal glands is 5 μm or less posterior to the base of the stylet. The vulva is close to the midpoint of the body; one gonad extends anteriorly and the other posteriorly. The cuticular annulations are well marked, slightly more than 1.5 μm apart and there are four lateral incisures. The tail is hemispherical or conoid, with annulations extending to the terminus. Males are slightly smaller than females. Bursae extend to the tail tip (2). *Rotylenchus* differs from *Scutellonema* by having a small phasmid, and from *Helicotylenchus* by the position of the dorsal esophageal gland orifice—close to the stylet knobs in *Rotylenchus* and at some distance in *Helicotylenchus*. See Fig. 6.15 for a diagram of a representative of this genus.

Biology

All spiral nematodes are migratory ectoparasites; occasionally specimens may enter roots completely. Populations of these nematodes may reach high levels in light sandy soils.

Pathology

Local lesions result from death of cells caused by feeding. General debilitation of roots results from multiple attacks, top growth is reduced, and leaves of some hosts become yellow.

Interaction with Other Pathogens

Fusarium oxysporum f. *pisi* is associated with *Rotylenchus robustus* in a problem of peas called "early yellowing."

Control

Application of nematicides controls spiral nematodes.

Research

A recent key to species has been published (1).

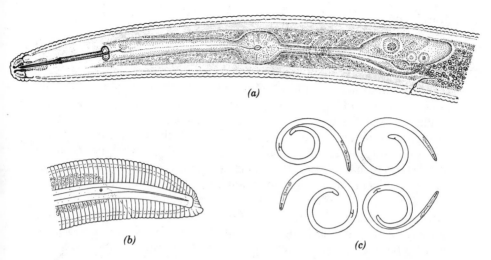

Fig. 6.15 *Rotylenchus buxophilus.* (a) Adult female, anterior portion; (b) female, posterior [(a) & (b) (×720)]; (c) typical shape of "spiral nematodes" after gentle heating (×54). (From M. R. Siddiqi. 1972. *Rotylenchus buxophilus.* Commonwealth Institute of Helminthology, Descriptions of Plant-parasitic Nematodes, Set 4, No. 55; see also *R. robustus*, Set 1, No. 11, © Commonwealth Agricultural Bureaux, by permission.)

References

1. Boag, B. and Hooper, D. J. 1981. *Rotylenchus ouensis* n. sp. (Nematoda: Hoplolaimidae) from the British Isles. Syst. Parasitol. 3:119 – 125.
2. Siddiqi, M. R. 1972. C.I.H. Descriptions of Plant-parasitic Nematodes Set 1, No. 11: *Rotylenchus robustus.* See also Set 4, No. 55: *R. buxophilus.*

HELICOTYLENCHUS

Introduction

Nematodes of this genus are common on many hosts, including hardwood trees, turf, soybean, olive, cotton, millet, tomatoes, and others. A species of this genus is an important pest of bananas.

Classification

Order Tylenchida; suborder Tylenchina; superfamily Tylenchoidea; family Hoplolaimidae; subfamily Rotylenchinae.

Morphology

Sexes are separate in some species and males are unknown in others. Females are from about 0.5 to 1 mm long, with a length/width ratio from 25 to 35. Males resemble females except for secondary sex characters. Lip region is hemispherical, not set off from the body, and lip annulations are not subdivided by longitudinal striations. Stylet is well developed, ranging in different species from 23 to 35 μm long. Esophageal glands partially surround the anterior end of the intestine, with the longest overlap ventrally. In most species, the dorsal esophageal gland opens behind the knobs at a distance equal to at least 1/4 length of the spear. The vulva is at about 60% of the length and there are an anterior and a posterior gonad. Phasmids are small, near the anus. Cuticular annulations are well marked, extending almost to the terminus. The female tail is usually more curved dorsally, with a hemispherical or ventrally elongated terminus. Bursae envelop the tails of males. See Fig. 6.16 for diagnostic morphological details.

Biology

These spiral nematodes, like the others, penetrate part way into roots. Occasionally the entire nematode enters a root to feed internally.

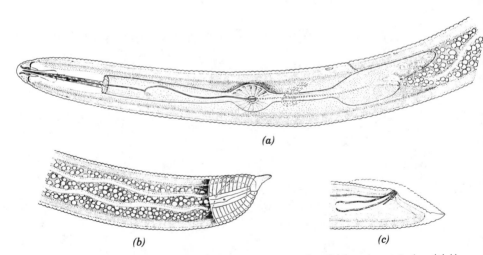

(a)

(b)   *(c)*

Fig. 6.16 *Helicotylenchus.* (a) *H. canalis,* female, anterior; (b) female, posterior; (c) *H. anhelicus* male, posterior. (a) and (b) (×705), (c) ×735. (From S. A Sher. 1966. Revision of the Hoplolaiminae (Nematoda) VI. *Helicotylenchus* Steiner, 1945. Nematologica 12: 1 – 56, by permission of E.J. Brill.)

Table 6.2 Lattice for the Separation of Hoplolaimus, Scutellonema, Rotylenchus, and Helicotylenchus[a]

	Not Subdivided	Subdivided	Large	Small	In Separate Parts of Body	Opposite, in Anal Region	With Forward-Pointing Cusps	Without Forward-Pointing Cusps	Close to Base of Spear	Far from Base of Spear
Head annulation	2, 4 *a*	1, 3 *b*								
Phasmid size			1, 2 *c*	3, 4 *d*						
Phasmid location					1 *e*	2, 3, 4				
Stylet knobs							1 *f*	2 – 4 *g*		
Location of duct opening of dorsal esophageal gland									1 – 3 *h*	4 *i*

[a] Numbers refer to the genera, letters refer to the parts of Fig. 6.17. Genera: 1. *Hoplolaimus*; 2. *Scutellonema*; 3. *Rotylenchus*; 4. *Helicotylenchus*.

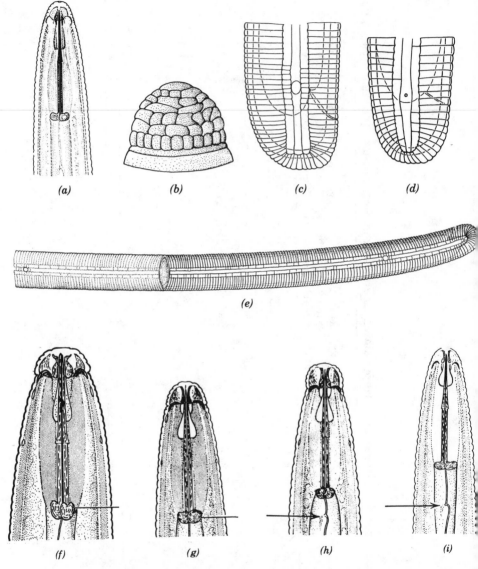

Fig. 6.17 Illustrations of characteristics mentioned in Table 6.2. **(a)** *Helicotylenchus dihystera*: head annulation not subdivided; **(b)** *Rotylenchus robustus*: head annulation subdivided; **(c)** *Scutellonema validum*: large phasmid; **(d)** *R. robustus*: small phasmid; **(e)** *Hoplolaimus californicus*: phasmids in separate parts of body; **(f)** *H. stephanus*: stylet knobs with forward pointing cusps; **(g)** *R. robustus*: stylet knobs with rounded cusps; **(h)** *R. robustus*: dorsal esophageal gland orifice close to base of stylet; **(i)** *Helicotylenchus anhelicus*: dorsal esophageal gland orifice far from base of spear.

Pathology

Local lesions in the cortex result from death of cells upon which the nematode feeds. Individual parasites destroy a few cells at a feeding site and are therefore mild pathogens. But when populations reach high levels, these ectoparasites severely damage roots.

Interaction with Other Pathogens

Bacterial wilt of Carnation caused by *Pseudomonas caryophylli* is more severe in the presence of *Helicotylenchus*. Nematodes alone do not induce wilt. In tomato wilt induced by *P. solanacearum*, *Helicotylenchus* increases severity and incidence of the disease.

Control

As in the case of infections of *Radopholus similis* on banana, underground root portions used for propagation are peeled and dipped in nematicides. These harbor low numbers of spiral nematodes and produce healthy plants. Genetic resistance has been found in cotton.

Research

A revision of the genus together with descriptions of new species including a key to species was published (5). The computer is being used as an aid in identification of species within large genera such as *Helicotylenchus* (2).

Helicotylenchus displays the same relationship to plants under stress as *Pratylenchus*. It reproduces better in roots of clipped bluegrass and banana than in intact plants (4,6).

Indian scientists are investigating the usefulness of oil cakes for nematode control. These are by-products of oil extraction from several species

---

(a) (From M. R. Siddiqi. 1972. C.I.H. Descriptions of Plant-parasitic Nematodes, Set 1, No. 9: *Helicotylenchus dihystera*, © Commonwealth Agricultural Bureaux, by permission.) (b), (d), (g) and (h) (From S. A. Sher. 1965. Revision of the Hoplolaiminae (Nematoda) V. *Rotylenchus* Filipjev, 1936. Nematologica 11:173 – 198.) (c) (From S. A. Sher. 1963. Revision of the Hoplolaiminae (Nematoda) III. *Scutellonema* Andrassy, 1958. Nematologica 9:421–443.) (e) and (f) (From S. A. Sher. 1963. Revision of the Hoplolaiminae (Nematoda) II. *Hoplolaimus* Daday, 1905 and *Aorolaimus* n. gen. Nematologica 9:267 – 295.) (i) (From S. A. Sher. 1966. Revision of the Hoplolaiminae (Nematoda) VI. *Helicotylenchus* Steiner, 1945. Nematologica 12: 1–56. (b – i) by permission of E. J. Brill.)

of plants. Substantial applications of oil cakes suppress populations of four species of phytonematodes on okra, including *Helicotylenchus* (3). Phenolics present in oil cakes appear to be the toxic agents for several kinds of phytonematodes (1). Two species of *Tagetes* depress populations of *Meloidogyne* and *Helicotylenchus*. The report gives no indication whether this effect results from the unsuitability of *Tagetes* as a host or from its nematicidal action (7).

Table 6.2 and Fig. 6.17 present the diagnostic characteristics of the four genera of "spiral nematodes".

References

1. Alam, M. M., Khan, A. M., and Saxena, S. K. 1979. Mechanism of control of plant parasitic nematodes as a result of the application of organic amendments to the soil. V. Role of phenolic compounds. Indian J. Nematol. 9:136 – 142.

2. Fortuner, R. and Wong, Y. 1984. Review of the genus *Helicotylenchus* Steiner, 1945. I: a computer program for identification of the species. Rev. Nématol. 7:385 – 392.

3. Khan, M. W., Khan, A. M., and Saxena, S. K. 1979. Suppression of phytophagus nematodes and certain fungi in the rhizosphere of okra due to oil-cake amendments. Acta Bot. Indica 7:51 – 56.

4. Mateille, T., Cadet, P., & Quénéhervé, P. 1984. Effect of pruning of the Poyo bananas on the development of *Radopholus similis* and *Helicotylenchus multicinctus* populations. Rev. Nématol. 7:355 – 361.

5. Sher, S. A. 1966. Revision of the Hoplolaiminae (Nematoda) VI. *Helicotylenchus* Steiner, 1945. Nematologica 12:1 – 56.

6. Stanton, N. L. 1983. The effect of clipping and phytophagous nematodes on net primary production of bluegrass, *Boutelous gracilis*. Oikos 40:249 – 257.

7. Vergel, G. A., Sierra, G. A. and Volcy, C. 1979. *Tagetes patula* and *T. erecta* for the control of *Meloidogyne incognita* and *Helicotylenchus dihystera*. Rev. Fac. Nac. Agron. Univ. Antioquia 32:65 – 71.

ROTYLENCHULUS

Introduction

In recent years, as modern agriculture replaces subsistence farming in the tropics, nematodes of this genus have assumed great importance. Very large populations of *Rotylenchulus reniformis* develop on susceptible hosts, including cotton, maize (in South Africa), tea, cowpea, bean, pineapple, soybeans, and sweet potato. *R. reniformis* is the best studied species of the genus.

Classification

Order Tylenchida; suborder Tylenchina; superfamily Tylenchoidea; family Hoplolaimidae; subfamily Rotylenchulinae.

Morphology

Some populations of this species include both sexes whereas others have only females. Adult females are obese, kidney-shaped, with raised vulva. Body length is up to 0.5 mm. Lip region is elevated, conoid, and not set off from the body; the labial framework is heavily sclerotized. Stylet length is 12 – 15 μm. The dorsal esophageal gland empties into the esophagus about one stylet length behind the knobs. This position is diagnostic for *Rotylenchulus*. The median bulb is large, glands are compact, partly overlapping the intestine. The vulva is located at about two-thirds of the body length; there are two gonads. Eggs are enormous, measuring 64 × 34 μm in a related species, *R. parvus*. Males of *R. reniformis* are degenerate, with poorly developed stylet and median bulb. They have a small bursa that does not extend to the tail tip (4). See Fig. 6.18.

Biology

Second-stage juveniles of *R. reniformis* emerge from eggs, do not feed, but undergo three molts in soil and develop into young adults. A stimulus from growing roots is required for the final molt. Young females enter partway into outer tissues of roots, swell, and deposit eggs into a gelatinous egg sac that often envelops the external portion of the female's body.

Pathology

Young females of *R. reniformis* penetrate the cortex to feed on a cell of the endodermis. This cell enlarges and a stimulus is transmitted to the outermost layer of pericycle cells. A curved sheet of 100 – 200 transformed cells results, extending laterally and longitudinally from the original site of cell penetration. These transformed cells have dense cytoplasm and have lost their central vacuole. In some hosts, cells of the pericycle tissue enlarge under the nematode's influence. In others, pericycle cells retain their usual size. No wall ingrowths such as those found in transformed cells of *Meloidogyne* infections are evident (Figs. 6.19, 6.20).

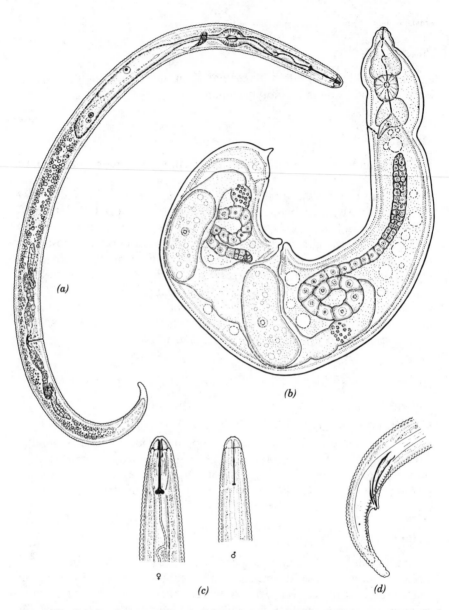

Fig. 6.18 *Rotylenchulus reniformis.* (**a**) Immature female (×570). Note overlapping glands, position of vulva, and dorsal esophageal gland opening far behind stylet. (**b**) Mature female (×347). Note two ovaries, anterior position of excretory pore, large metacorpus and glands. (**c**) Anterior ends of female and male (×890). (**d**) Posterior end of male (×890). Note weak stylet of male and small caudal alae. (From M. R. Siddiqi. 1972. Commonwealth Institute of Helminthology Descriptions of Plant-parasitic Nematodes, Set 1, No. 5: *Rotylenchulus reniformis*; see also Set 6, No. 83: *R. parvus.* © Commonwealth Agricultural Bureaux, by permission.)

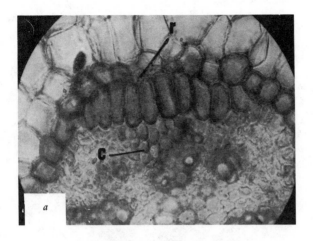

(b)

Fig. 6.19 Histopathology of plants with *Rotylenchulus reniformis*. (a) Cross section of cotton root showing transformed cells of the pericycle (r) and some affected cells in phloem region (c); (b) Tangential section through a sheet of altered cells in soybean (×280), nucleus of affected cell (N), section of nematode (NE). (a) (From E. Cohn. 1973. Histology of the feeding site of *Rotylenchulus reniformis*. Nematologica 19:455 – 458, by permission of E.J. Brill.) (b) (From M. G. K. Jones and V. H. Dropkin. 1975. Cellular alterations induced in soybean roots by three endoparasitic nematodes. Physiol. Plant Pathol. 5:119 – 124, by permission of Academic Press.)

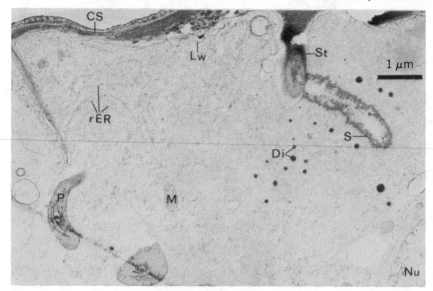

Fig. 6.20 Electronmicrograph of endodermal cell of soybean penetrated by stylet of *R. reniformis* (×13600). Cell is identified by Casperian strip. Wall near stylet is lysing two days after inoculation whereas separated vestigial wall in advanced stage of lysis separates original pericycle cell from endodermal cell. The plastid suggests that cytoplasmic flow is away from stylet. (CS) Casperian strip; (Di) dense spherical inclusions; (Lw) lysing wall; (M) mitochondrion; (Nu) nucleus; (P) plastid; (rER) rough endoplasmic reticulum; (S) secretion or exudate; (St) stylet. (From R. V. Rebois, P. A. Madden, and B. J. Eldridge. 1975. Some ultrastructural changes induced in resistant and susceptible soybean roots following infection by *Rotylenchulus reniformis*. J. Nematol. 7:122 – 139, with permission.)

    R. macrodoratus induces a single, very large giant cell in soybean endodermis or pericycle. The giant cell replaces part of the stele. Its nucleus is relatively enormous and has an irregular shape. Similar single, uninucleate giant cells develop in infections with several other sessile nematodes of different genera.

Interaction with Other Root-Inhabiting Microorganisms

Infection with *R. reniformis* enhances the action of pathogenic fungi. Wilt of okra caused by *Rhizoctonia solani* occurs earlier in the presence of this nematode than in its absence. *Verticillium* and *Fusarium* wilts of cotton are also increased by simultaneous infection with *Rotylenchulus*.

    However, *Glomus fasciculatus*, the mycorrhizal fungus, inhibits reniform nematodes at several stages of the life cycle on tomato. And on the legume cowpea, *R. reniformis* at levels sufficient to inflict damage do not inhibit nodulation.

Control

Under favorable conditions, this nematode can survive fallow for extended periods. Because of the wide host range, including common weeds, clean cultivation is important for control. Nematicides are commonly used and genetically resistant cultivars are available in several crops.

Research

As knowledge of phytonematodes advances, we are becoming aware of the great diversity that exists both at the specific and subspecific levels. Morphologically and physiologically distinct groups of *Rotylenchulus* occur in Japan. Reproduction without males is characteristic of some, but not of all populations (2). Studies of behavioral responses in *Rotylenchulus* and two other genera of phytonematodes show that each species has its own array of sensory perceptions for finding food and mates (3).

The role of mycorrhizal fungi in relation to nematodes has interested several investigators. These symbiotic fungi increase absorption of minerals, especially phosphorus, thereby improving plant growth. In addition, the fungi may protect roots from attack by nematodes (5).

Details of histopathology in plants infected with nematodes continue to be examined. A comprehensive review comparing histopathologies induced by endoparasitic phytonematodes, including *Rotylenchulus* is presented in Ref. (1).

References

1. Jones, M. G. K. 1981. The development and function of plant cells modified by endo-parasitic nematodes. In: Plant Parasitic Nematodes, Vol. 3 (B. M. Zuckerman, and R. Rohde, eds.). Academic Press, New York, pp. 255 – 279.
2. Nakasono, K. 1983. Studies on the morphological and physio-ecological variations of the reniform nematode *Rotylenchulus reniformis* Linford and Oliviera, 1940 with an emphasis on differential geographical distribution of amphimictic and parthenogenetic populations in Japan. Bull. Nat. Inst. Agri. Sci., Ser. C 38:1 – 67.
3. Riddle, D. L. and Bird, A. F. 1985. Responses of the plant parasitic nematodes *Rotylenchulus reniformis*, *Anguina agrostis*, and *Meloidogyne javanica* to chemical attractants. Parasitology 91:185 – 195.
4. Siddiqi, M. R. 1972. Commonwealth Institute of Helminthology, Descriptions of Plant-parasitic Nematodes, Set 1, No. 5: *Rotylenchulus reniformis*; see also Set 6, No. 83: *R. parvus*.

5. Sitaramaiah, K. and Sikora, R. A. 1982. Effect of the mycorrhizal fungus *Glomus fasciculatus* on the host-parasite relationship of *Rotylenchulus reniformis* in tomato. Nematologica 28:412 – 419.

BELONOLAIMUS

Introduction

These elongate, voracious nematodes with long stylets are among the most damaging ectoparasites known. Populations increase to damaging levels on a wide variety of hosts including small grains, forage, fruits, ornamentals, trees, turf and weed grasses, soybeans, cotton, and corn. In sandy soils they sometimes completely destroy a crop.

Classification

Order Tylenchida; suborder Tylenchina; family Belonolaimidae; subfamily Belonolaiminae.

Morphology

Sexes are separate. Females of *B. longicaudatus* are 2 – 3 mm long by about 50 μm wide. Lip region is hemispherical, offset, and divided into four principal lobes plus two smaller ones. The thin, flexible spear has knobs and is 100 – 140 μm long. Lumen of the esophagus coils within the procorpus when the spear is retracted and straightens when it protrudes. Esophageal glands overlap the intestine. Two opposed, outstretched gonads join at the midbody vulva. The tail is about five times as long as the anal body width. Males are similar to females in size and appearance. Bursae extend to the tail tip (3). Figure 6.21 illustrates the morphology.

Biology

B. longicaudatus prospers in warm sandy soils and causes damage especially in southeastern United States. Isolated infestations are known elsewhere. Another species occurs as far north as the Nebraska – South Dakota border. Populations of *B. longicaudatus* from Georgia differ from those in North Carolina in several characteristics including host range, optimal temperature for reproduction, and virulence. Perhaps the populations represent separate species because there are slight morphological differences and matings between them result in infertile offspring.

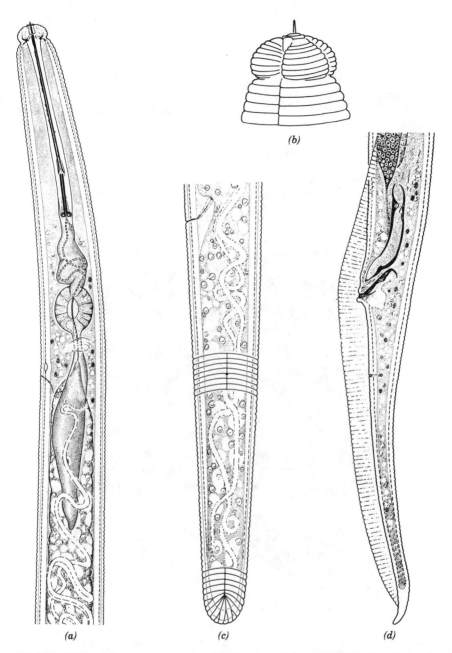

(b)

(a)   (c)   (d)

Fig. 6.21 *Belonolaimus longicaudatus.* (a) Female anterior (×500). (b) Female head (×884). (c) Female posterior (×500). (d) Male posterior (×500). (From K. J. Orton Williams. 1974. Commonwealth Institute of Helminthology, Descriptions of Plant-parasitic Nematodes, Set 3, No. 40: *Belonolaimus longicaudatus,* © Commonwealth Agricultural Bureaux, by permission.)

Pathology

Belonolaimus nematodes destroy cells in outer portions of roots and produce necrotic lesions. Root growth is severely restricted when nematode populations are high; consequently top growth and yield are reduced (Fig. 6.22).

Interaction with Other Pathogens

Both nematodes and pathogenic fungi act together to cause great damage to roots. However, this nematode, in contrast to *Heterodera* and *Meloidogyne*, does not markedly disturb nodule formation by nitrogen-fixing bacteria.

Fig. 6.22 Effect of *Belonolaimus* on growth of carrot. (a) Left, no nematodes; right, approximately 5000 nematodes present per 20-cm pot at time of planting. (b) Effect of nematicide on stand and growth of carrot in *Belonolaimus*-infested soil. Left, aldicarb treatment; right, no treatment. (From H. L. Rhoades. 1975. Pathogenicity and control of the sting nematode, *Belonolaimus longicaudatus*, on carrot. Plant Dis. Rep. 59: 1021 – 1024, by permission.)

Control

Because the host range is broad, rotations to reduce damaging populations must be based on knowledge of the local nematodes and their reproduction on proposed alternate crops. Moreover, some of the best candidates for rotation have little economic value and are therefore not suitable for profitable farming (e.g., hairy indigo). Weed control is also necessary to avoid increase of inoculum. Sources of resistant grasses suitable for forage or cover crops are being found and these may eventually be incorporated into useful cultivars.

Nematicides are the primary control for economically important levels of *Belonolaimus*. Spectacular results have been reported, especially after incorporation into soil of granular compounds such as aldicarb, Dasanit, and carbofuran. Soil amendments with organic crop residues or composted sewage sludge also have proven useful.

Research

The existence of geographic variants is well documented (1, 4). Taxonomic status of these populations is not completely settled because there is some restriction of the flow of genes among populations (6). *B. longicaudatus* appears to be more sensitive to ecological factors than some other nematodes. For example, reproduction is possible only in soils with high proportions of sand. Small increases in proportions of clay can inhibit population growth on favorable hosts (5). Quality and intensity of light in greenhouse cultures affect population increase of these nematodes on cotton, probably through an effect on host physiology (2). In growth chambers, however, the influence of light on this nematode feeding on soybeans is not marked.

It is always a problem to determine which nematodes among the complex of species present are responsible for the observed poor growth and yield of a crop. One approach is to test each species separately in pot experiments. But results in the greenhouse may bear little relation to events in the field. An interesting analysis of this complex situation is illustrated in Ref. (7). Plots were established in fields infested with several known phytonematodes. Different levels of populations were achieved by various treatments with nematicides. By careful measurement of nematode populations and growth characteristics of the crop, a matrix of correlations between nematode numbers and crop performance was established. In general, yield fell as nematode counts rose. *Belonolaimus* was most damaging, *Criconemoides* next, and *Meloidogyne* least.

References

1. Abu-Gharbieh, W. I. and Perry, V. G. 1970. Host differences among Florida Populations of *Belonolaimus longicaudatus* Rau. J. Nematol. 2:209 – 216.
2. Barker, K. R., Hussey, R. S., and Yang, H. 1975. Effects of light intensity and quality on reproduction of plant parasitic nematodes. J. Nematol. 7:364–368.
3. Orton Williams, K. J. 1974. Commonwealth Institute of Helminthology, Descriptions of Plant-parasitic Nematodes, Set 3, No. 40: *Belonolaimus longicaudatus*.
4. Robbins, R. T. and Barker, K. R. 1973. Comparisons of host range and reproduction among populations of *Belonolaimus longicaudatus* from North Carolina and Georgia. Pl. Dis. Rep. 57:750 – 754.
5. Robbins, R. T. and Barker, K. R. 1974. The effects of soil type, particle size, temperature, and moisture on reproduction of *Belonolaimus longicaudatus*. J. Nematol. 6:1 – 6.
6. Robbins, R. T. and Hirschmann, H. 1974. Variation among populations of *Belonolaimus longicaudatus*. J. Nematol. 6:87 – 94.
7. Sasser, J. N., Barker, K. R. and Nelson, L. A. 1975. Correlations of field populations of nematodes with crop growth responses for determining relative involvement of species. J. Nematol. 7:193 – 198.

HETERODERA AND OTHER CYST NEMATODES

Introduction

The name "cyst nematode" refers to the female's swollen body whose thick cuticle hardens at death and remains in soil as a cyst containing embryonated eggs. Problems in Europe caused by cyst nematodes on beets during the mid-nineteenth century and on potatoes during World War II stimulated interest in phytonematology. The most important species are *H. schachtii* on beets, cabbage, and related crops; *H. avenae* on oats, barley, and wheat; *H. glycines* on soybeans; and *Globodera pallida* and *G. rostochiensis* on potato.

Classification

Order Tylenchida; suborder Tylenchina; superfamily Heteroderoidea; family Heteroderidae; subfamily Heteroderinae.

Morphology

This account applies to the genus *Heterodera*. Sexes are separate in some species and males are lacking in others. Adult females of *Heterodera* spp. are lemon-shaped, with an anterior neck and a posterior vulval cone. Size ranges from 300 to 1000 μm long. Width is more than half the length. Cyst sizes are variable; those in some species tend to be larger than in others.

Moreover, cysts on a single plant may also be variable. Lips are annulated and are not offset from the short neck that leads into the swollen body. The stylet is about 25 μm long, and has small knobs. There is no tail; the terminus remains as a point on the vulval cone. The median bulb is spherical and esophageal glands are prominent. A double uterus containing 200 – 500 or more eggs almost fills the body cavity at maturity. In some species, the female secretes a gelatinous sac into which some eggs are deposited. The adult cuticle wall is thick. When the female dies, this wall hardens, darkens, and is marked by a reticulate or zigzag pattern. The vulva is a slit in a complex organ. Taxonomists distinguish species by differences in the anatomy of this structure (12). Figures 6.23 and 6.24 show cysts of two species of this genus.

Second-stage juveniles are elongate, active nematodes, measuring between 400 and 500 μm long with a length/width ratio between 21 and 26. The head is hemispherical, annulated, and offset, and is equipped with a strong cephalic skeleton. The sturdy stylet has prominent knobs. Esophageal glands are long and overlap the intestine. The cuticle has distinct lateral fields, usually with four incisures. The tail has a clear terminal portion that ends in a point or narrow rounded tip.

Males are elongate and motile. They vary in length even in the same species, usually from 1 to 1.5 mm, with a length/width ratio of 32 – 51. The rounded, offset head has annulated lips and a heavily sclerotized cephalic skeleton. The stylet varies from 23 to 30 μm in length depending

Fig. 6.23 Cysts of *Heterodera graminophila* (× 15). White cysts are young, dark cysts are older. (From A. M. Golden and W. Birchfield. 1972. *Heterodera graminophila* n. sp. (Nematoda:Heteroderidae) from grass with a key to closely related species. J. Nematol. 4:147 – 154, by permission.)

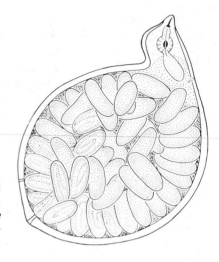

Fig. 6.24 Cyst of *Heterodera betulae* filled with eggs (× 114). (From H. Hirschmann, and R. D. Riggs. 1969. *Heterodera betulae* n. sp. (Heteroderidae), a cyst-forming nematode from river birch. J. Nematol. 1: 169 – 179, by permission.)

⊢—————————⊣ 200 μm

on the species. The median bulb is oval and long. The posterior body is usually twisted 90° or more. Lateral fields have four incisures. The bluntly rounded tail has a pair of curved spicules close to its terminus and is without caudal alae.

In addition to *Heterodera* there are other genera of cyst nematodes.

Globodera is distinguished by its globular shape in contrast to the lemon shape of *Heterodera*

Punctodera is elongate – ovoid with a protruding neck and no vulval cone.

Sarisodera differs from *Heterodera* in the presence of hypertrophied vulval lips and the absence of various structures associated with the vulva in *Heterodera* and *Punctodera*. Males of *Sarisodera* have straight spicules situated at the very tip of the pointed tail.

More genera are being reported, differing in one or a few characteristics from the ones listed (5).

Biology

Cyst nematodes are specialized endoparasites adapted for long survival in soil in the absence of host plants. Eggs are retained in the uterus except that in a few species some eggs are deposited in a mucoid sac secreted by the adult female. Juveniles molt once within eggs before hatching. When the female dies, embryonated eggs are retained within the body. A portion of these juveniles hatch from time to time. Unhatched juveniles of some species remain quiescent within cysts until a stimulus from growing roots reaches them. Then juveniles become active, cut their way out of the eggshell, leave cysts through breaks in the wall, and migrate to growing roots. A few embryonated eggs may remain viable for as long as 15 years in the absence of host plants!

Once inside a root, the juvenile moves for a few days, then finds a location close to the vascular system. Here it induces formation of a feeding site with altered cells upon which it feeds for the rest of its life cycle. J–2s swell and molt three times to the adult stage. All developmental stages feed. While a female grows, tissues surrounding the swelling nematode disintegrate and the female emerges to the root surface, except for the neck portion, which remains in feeding position. Upon death, cysts are easily dislodged from roots and remain in the soil. Males complete their development in time to mate with newly formed adult females exposed on the root surface.

There are many species of cyst nematodes, in both tropical and temperate zones. A key to species of the Western hemisphere is presented in Ref. (12).

Pathology

Infective juveniles leave a trail of destruction as they move through root cortex en route to a feeding site close to or in the stele. Here cells enlarge and undergo partial dissolution of their walls so that a **syncytium** of interconnected cells is formed, shown in Fig. 6.25. These cells exhibit signs of active metabolism: an enlarged nucleus, dense cytoplasm, and increased numbers of organelles. The nematode feeds in one position, withdrawing cell contents at one end of the syncytium. Cell wall ingrowths are evident, indicating active movement of solutes into a syncytium. Galls are not formed except in a few host – Heterodera combinations. Cell modifications induced by these nematodes illustrate the exquisite adjustments of host and parasite to each other by which a parasite induces its host to assemble a specialized structure upon which the nematode feeds as it grows and develops.

Roots with large numbers of cyst nematodes are poorly developed, plants are under moisture stress, and nutrient balance is disturbed. Nitrogen fixation by bacteria is inhibited. Because viable eggs persist for long periods in soil, populations of cyst nematodes build to high levels with

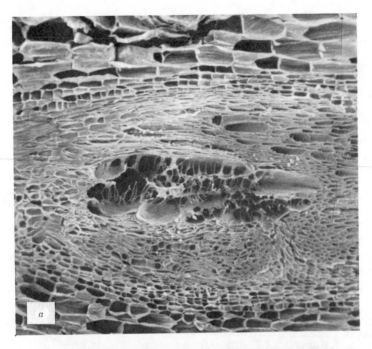

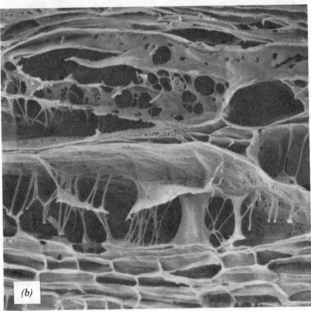

Fig. 6.25 Cell walls of syncytium induced by *Heterodera glycines* in soybean roots. (a) Overall view of syncytial area, longitudinal section ($\times$ 100); (b) walls showing holes and columns that suggest that the cell has enlarged, stretching the walls ($\times$ 300). Scanning electron micrographs by William Nace.

repeated plantings of susceptible hosts. The first indication of a problem is often the presence of a small area in a field where the crop grows poorly, and weeds are numerous. Each year the area enlarges until complete crop failure may result.

Interaction with Other Pathogens

This does not seem to be an important factor in the pathology of roots invaded by *Heterodera*.

Control

Growers cannot always afford long rotations of nonhosts for control of cyst nematodes. However, where cyst nematodes cause serious problems, rotations may be enforced. Potato production in the Netherlands is tightly regulated. Government inspectors count cysts of *G. rostochiensis* and *G. pallida* and the counts are used to determine permissible farming practices: length of rotations, resistant varieties in the rotations, and nematicide use. Much attention is devoted to production of genetically resistant cultivars. Because nematode species have multiple alleles of genes for virulence, wide use of a resistant cultivar selects out nematodes capable of reproducing on it. This applies especially to crops such as soybeans grown in warm soil. Breeding for resistance therefore must continue. This subject is discussed in Chapter 9.

Early planting to permit a crop to begin growth before nematodes become active is effective with beets grown in western United States.

Research

Cyst nematodes are highly specialized parasites well adapted to survive during absence of suitable hosts. Once the existence of hatching factors in potato root exudates was demonstrated, nematologists sought to understand how the quiescent juvenile coiled inside an egg within a cyst in soil becomes aware that roots of a host are growing nearby. What kind of stimulus reaches this nematode and how does it act? During the 1940s and 1950s some of the best chemists and nematologists of Europe tried in vain to find answers for potato cyst nematodes. They hoped to produce an artificial hatching agent for application in the field when hosts were not present and so to exhaust food reserves of infective juveniles before the host crop became available. However, the potato and nematode biologies were obstacles. It was very difficult to get reliable quantitative measures of the degree of stimulation for use in identification of the active components in exudates. Compounds extracted from roots were unstable and subject to biological degradation. A team of Japanese investigators recently announced the chemical structure of a hatching stimulant from kidney bean roots that is effective for *H. glycines* at extremely low concentrations (6). In addition to natural hatching factors, a number of chemical agents such as picric acid and synthetic dyes are known hatching stimulants for some cyst nematode species.

How do these agents work? One hypothesis under study is that the stimulus changes permeability of the eggshell, permitting water to enter and solutes surrounding the egg to leak out (13). As the osmotic concentration of fluids surrounding the coiled nematode declines, the 2nd stage juvenile becomes active and cuts its way out of the shell.

The histopathology of infection was described in a series of papers by Jones (9). He recognized that altered cells at feeding sites display attributes of "transfer" cells known in healthy plants. These cells have enlarged nuclei, numerous organelles, and most particularly, fingerlike, inwardly directed projections of their walls (wall ingrowths). These ingrowths greatly increase surface areas of the plasma membrane. Because ingrowths are adjacent to vascular elements, transfer cells are presumed to enhance transport of solutes to places with high utilization of nutrient, such as at the base of a growing leaf, or in the ovary next to a growing embryo. Sessile nematodes require compounds to support growth and egg production; the syncytium plus nematode constitute a "metabolic sink" to which nutrients flow from the rest of the plant.

Intensive efforts to develop genetically resistant cultivars of potato, beets, oats and barley, soybeans, and other crops have continued for many years. Resistant cultivars, however, have a limited useful life. Selection pressures imposed on nematode populations by widespread cultivation of resistant plants alter the genetics of parasite populations. There are always a few individuals that reproduce on resistant host plants. Successful nematodes increase, and after years of cultivation, resistant cultivars are no longer useful. It is obviously important to keep track of changes in nematode populations. In recent years several investigators are exploring techniques of gel electrophoresis to distinguish nematodes by the movement of soluble proteins in electric fields. This is a well established technique for characterizing proteins. The method is based on the assumption that various populations of nematodes have distinctive arrays of proteins by which they can be differentiated. Both enzymes and nonenzymatic proteins can be visualized on the gels (1, 4).

The influence on host plant physiology of attack by cyst nematodes has stimulated research to find practical methods of minimizing crop damage. Growers have known for many years that potassium nutrition is of great importance. Calcium translocation from roots of oats is greater in the presence of *H. avenae* than in healthy plants. The authors attribute this to disruption of the endodermis (14). X-ray microanalysis demonstrates that concentration of a number of minerals in vacuoles of affected sugar beets is increased whereas concentration of potassium is depressed (2). Addition of minerals to soil may therefore alleviate crop damage. Moreover, nitrogen fertilizer in various forms, especially urea, drastically reduces larval hatch (7).

Cyst nematodes depress formation of nodules by nitrogen-fixing bacteria on roots of legumes. Some populations of *H. glycines* are more effective than others in this respect. Nematodes on one-half of a split root of a soybean plant can inhibit nodulation on the noninfected part of the root system. This suggests a change in the physiology of the entire plant (10). Leghemoglobin is an essential compound for nitrogen fixation; concentrations of this substance are reduced in nodules of soybeans infected with *H. glycines* (8). Iron and carbohydrate metabolism is also disturbed in nodules on infected roots (11).

A new direction of research for chemical control is illustrated in Ref. (3). What compounds are synthesized by nematodes and not by plants? Can these be blocked by known inhibitors of the synthesis? Collagen is a constituent of the cuticle, and chitin is present in eggshells, but these polymers do not occur in plants. Prevention of collagen and chitin synthesis might affect nematodes without harming plant hosts. And indeed the idea has merit. Inhibitors of collagen and chitin formation protect plants with cyst nematodes. This approach is promising.

Summary

Research on cyst nematodes has been stimulated by the great economic importance of these pests. Whereas most investigations into the biology, pathology, and control of many other genera are descriptive in nature, work on cyst and root-knot nematodes employs techniques of sophisticated contemporary biology. Detailed biochemistry, identification of proteins by two-dimensional gel electrophoresis, chemical identification of hatching agents, X-ray microanalysis of cell contents, use of enzyme inhibitors, and beginning use of modern gene manipulation all show that phytonematologists are keeping abreast with the exciting advances in biology. Attention to cyst nematodes has served in the past to elevate phytonematodes to a position of importance in agriculture. Now researchers on these nematodes are showing that understanding of the intricate interaction of plants and nematodes and of nematodes and their environment will come only with the use of the best available science.

References

1. Bakker, J. 1987. Protein variation in cyst nematodes. Ph.D. Thesis published by Dept. of Nematology, Agricultural Univ., Binnenhaven 10, Wageningen, The Netherlands, 159 pp.

2. Barth, P., Stelzer, R. and Wyss, U. 1983. Changes in the inorganic metabolism in sugarbeet infected with *Heterodera schachtii*. Kali-Briefe 16:627 – 638.

3. Evans, K. 1984. An approach to control of *Globodera rostochiensis* using inhibitors of collagen and chitin synthesis. Nematologica 30:247 – 250.

4. Fleming, C. C. and Marks, R. J. 1983. The identification of the potato cyst nematodes *Globodera rostochiensis* and *G. pallida* by isoelectric focusing of proteins on polyacrylamide gels. Ann. Appl. Biol. 103:277 – 281.

5. Franklin, M. T. 1972. Commonwealth Institute of Helminthology, Descriptions of Plantparasitic Nematodes, Set 1 No. 1: *Heterodera schachtii*; see also Set 1, No. 2: *H. avenae*; Set 4, No. 46: *H. trifolii;* Set 4, No. 47: *H. goettingiana*; Set 4, No. 48: *H. sacchari*; Set 5, No. 61: *H. carotae*; Set 6, No. 90: *H. cruciferae*.

6. Fukazawa, A., Furusaki, A., Ikura, M. & Masumune, T. 1985. Glycinoeclepin A, a natural hatching stimulus for the soybean cyst nematode. J. Chem. Soc., Chem. Commun.:222 – 224.

7. Grosse, E. and Decker, H. 1984. Studies on the influence of fertilizer nitrogen on larval hatching and population dynamics of *Heterodera avenae*. Arch. Phytopathol. Pflanzenschutz 20:135 – 143.

8. Huang, J. S. and Barker, K. R. 1983. Influence of *Heterodera glycines* on leghemoglobins of soybean nodules. Phytopathology 73:1002 – 1004.

9. Jones, M. G. K. 1981. The development and function of plant cells modified by endoparasitic nematodes. In: Plant Parasitic Nematodes, Vol. 3, (B. M. Zuckerman and R. A. Rohde, eds.) Academic Press, New York, pp. 255–279.

10. Ko, M. P., Barker, K. R. and Huang, J. S. 1984. Nodulation of soybeans as affected by half-root infection with *Heterodera glycines*. J. Nematol. 16:97 – 105.

11. Ko, M., Huang, P., Huang, J. S., and Barker, K. R. 1985. Accumulation of phytoferritin and starch granules in developing nodules of soybean roots infected with *Heterodera glycines*. Phytopathology 75: 159 – 164.

12. Mulvey, R. H. & Golden, A. M. 1983. An illustrated key to the cyst-forming genera and species of Heteroderidae in the western hemisphere with species morphometrics and distribution. J. Nematol. 15:1 – 59.

13. Perry, R. N., Clarke, A. J., Hennessy, J., and Beane, J. 1983. The role of trehalose in the hatching mechanism of *Heterodera goettingiana*. Nematologica 29:323 – 334.

14. Price, N. S. and Sanderson, J. 1984. The translocation of calcium from oat roots infected by the cereal cyst nematode *Heterodera avenae* (Woll.) Rev. Nématol. 7:239 – 243.

MELOIDOGYNE

Introduction

This genus includes the most important species of phytonematodes in agriculture over the world. The name "root-knot nematodes" refers to the characteristic galls associated with these nematodes. Hosts of *Meloidogyne* spp. include vegetables, row crops, fruit trees, and weeds. The genus is especially important in tropical agriculture.

Classification

Order Tylenchida; suborder Tylenchina; superfamily Heteroderoidea; family Meloidogynidae.

Morphology

Adult females are flask-shaped sedentary endoparasites with a short neck and no tail. They measure more than 0.5 mm long by 0.3 – 0.4 mm wide. Lip region is small and bears up to three annules. A delicate stylet, 12 – 15 μm long, curves dorsally, and has prominent basal knobs. The female esophagus contains a prominent spherical metacorpus with large crescentic plates. Esophageal glands are large, compact, and close to the metacorpus and overlap the intestine. The dorsal esophageal gland duct is enlarged into an ampulla just behind its junction with the lumen of the esophagus. The excretory canal opens to the exterior far forward, sometimes just posterior to the stylet base. The intestine loses its form in the adult female and is not connected to the rectum. Uteri of the twin gonads join just anterior to the vulva. Eggs are deposited into an external egg sac secreted by rectal gland cells. The female cuticle in some species may reach 30 μm in thickness. A distinctive pattern of striations surrounding vulva and anus (**perineal pattern**) is used for species identification.

Juveniles resemble those of *Heterodera* except that they are more delicate, with a shorter and thinner stylet.

Adult males are elongate, and move about slowly in soil. They vary in length, with a maximum up to 2 mm, and with a length/width ratio close to 45. The head is not offset, and the stylet is almost twice as long as that of the female. The male tail is short and rounded, and the posterior part of the body is twisted, as much as 180°. Males have one or two testes. Intersex males are common in some species (12).

Figure 6.26 shows head structures of a second-stage juvenile and Fig. 6.27 is a diagram of adult female anatomy.

Biology

These obligate endoparasites are widely distributed in both tropical and temperate climates. Reproduction without males is the rule in many spe-

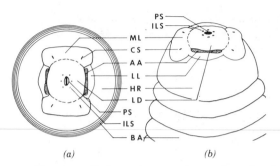

(a)   *(b)*

Fig. 6.26 Diagram of head morphology of a second-stage juvenile of the genus *Meloidogyne*. (**a**) Face view. (**b**) Lateral view. (AA) Amphidial aperture; (BA) body annule; (CS) cephalic sensillum; (HR) head region; (ILS) inner labial sensillum; (LD) labial disc; (LL) lateral lip; (ML) medial lip; (PS) prestoma. (From J. D. Eisenback and H. Hirschmann. 1979. Morphological comparison of second-stage juveniles of six populations of *Meloidogyne hapla* by SEM. J. Nematol. 11:5 – 16, by permission.)

cies, but in others both sexes are necessary. The role of male intersexes is unknown. Eggs are deposited into a gelatinous egg sac that probably protects them from desiccation and perhaps from microorganisms. In most host – parasite combinations a gall is formed from which an egg sac usually protrudes. Freshly deposited egg sacs are colorless and become brown upon aging. Eggs contain single-celled zygotes when deposited. The embryo develops into a juvenile that molts once within the egg. Second- stage juveniles hatch under favorable temperature and moisture conditions and move through soil toward growing root tips. They penetrate, usually in the region of root elongation, by breaking into a cell with repeated jabs of the stylet. Once within a root, a juvenile moves between cells until it locates a site close to the stele, often in the area of a developing side root. Here it becomes sedentary, and causes the transformation of cells upon which it feeds. The juvenile swells, molts rapidly a second and third time without feeding, and matures into an adult male or female. Adult males elongate within the fourth-stage cuticle and emerge from roots. Several males may aggregate within a single egg sac. An adult female remains attached to its feeding site within the stele and with its posterior at the root surface. It continues to produce eggs throughout life, sometimes reaching a total of more than 1000 eggs. The life cycle varies in length according to both host and temperature. It may be as short as 3 weeks and as long as several months. The sex ratio is also influenced by the environment. Males are more numerous when roots are heavily attacked or nutrition is inadequate. Although root exudates enhance hatching, they are not required for successful life cycles.

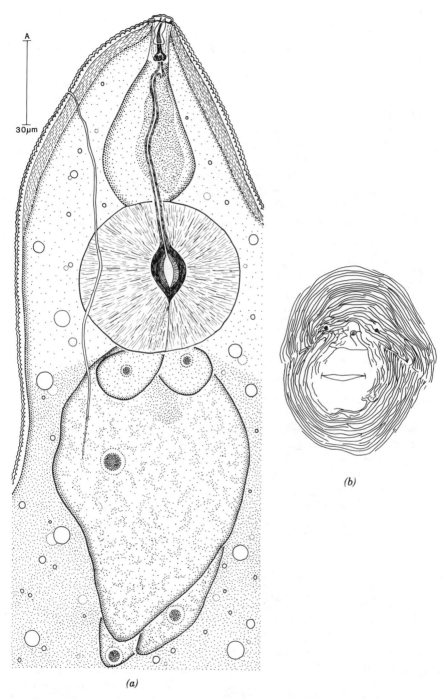

(a)

(b)

Fig. 6.27 Line drawings of females of *Meloidogyne arenaria*. (a) Esophageal region, lateral (×740); (b) perineal pattern (×445). (From G. M. Cliff and H. Hirschmann. 1985. Evaluation of morphological variability in *Meloidogyne arenaria*. J. Nematol.17:445 – 459, by permission.)

Infective juveniles contain great amounts of stored lipids. During starvation, these are utilized, and as seen through the microscope, the intestine appears to be banded. Some lipid-filled cells are opaque whereas others are clear, having lost their stored lipid. During migration through soil, juveniles utilize their food reserves and eventually die after a few months without a host. Nevertheless, soil free of host plants may contain infective larvae for up to a year. A few nematodes probably find niches where they survive with low rates of metabolism. Perhaps these niches are in the centers of soil crumbs where the nematodes are protected from desiccation and remain inactive in an atmosphere of low oxygen tension (10).

Species of *Meloidogyne* have large host ranges including weeds and cultivated plants. As in all well-studied species of phytonematodes, there are distinct subspecific populations (races or pathotypes) with distinctive host ranges. Protein separations by electrophoresis as well as host range tests are used to differentiate these populations of *Meloidogyne*.

Pathology

Infection with *Meloidogyne* initiates a series of events that alters the entire physiology of a host plant. An individual nematode has only a minor effect, but the cumulative result of large numbers is severely damaging. Two distinct pathologies occur:

(a) Cells of cortex and pericycle close to juveniles enlarge and divide. Root galls result. Second generation juveniles emerging from eggs at the surface of a gall often reinvade the nearby root. Each juvenile stimulates an increment of enlargement so that large galls the size of a walnut develop in heavy infections of certain hosts. Both the nematode and newly formed host tissue act as metabolic sinks to which the plant diverts nutrients normally sent to growing leaves, flowers, and fruits. Figure 6.28 shows a spectacular example of galled bean roots.

(b) Infective juveniles penetrate through the endodermis to reach the stele. About five to seven cells immediately surrounding the nematode's head enlarge to become specialized **giant cells**. These are the source of its nutrition throughout development to maturity. Giant cells are much larger than others, their nuclei also enlarge, become polyploid, and undergo a series of synchronous divisions. Hundreds of nuclei may be present in a single giant cell. Organelles also increase, the central vacuole breaks up into many smaller vacuoles, and ingrowths develop on walls bordering vascular elements. See Fig. 6.29. Giant cells are considered to be enlarged versions of "transfer" cells. When the nematode dies, giant cells disintegrate and parenchyma replaces them.

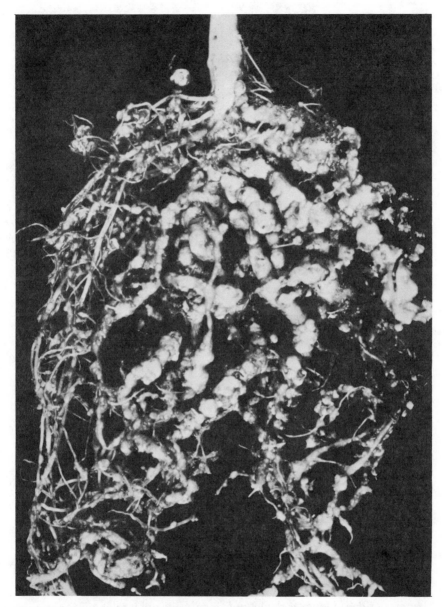

Fig. 6.28 *Meloidogyne* galls on bean roots. (From A. L. Taylor and J. N. Sasser. 1978. Biology, Identification, and Control of Root-knot Nematodes, *Meloidogyne* species. North Carolina Univ. Graphics, Raleigh, 111 pp., by permission.)

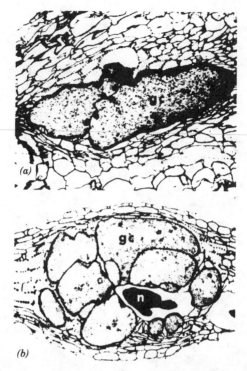

Fig. 6.29 Giant cells induced by *Meloidogyne incognita*. (a) Longitudinal section through a whole gall in a coleus root. Two giant cells (gc) are filled with cytoplasm, some vascular elements occur outside these cells, and vacuolate parenchyma cells of the gall surround the vascular and giant cells. The dark staining area of the giant cell walls are composed of many fine, branching transfer cell wall ingrowths (×160). (b) Longitudinal section of giant cells (gc) in the stele of a dwarf balsam root from which the cortical and gall parenchyma cells have been removed. The head of the nematode (n) has stained intensely. (×116) (From M. G. K. Jones, A. Novacky, and V. H. Dropkin.1975. Transmembrane potentials of parenchyma cells and nematode-induced transfer cells. Protoplasma 85:15 – 37, by permission of Springer-Verlag, New York.)

As studies of Heteroderidae and Meloidogynidae have ranged more widely, an array of variations has come to light. All species of *Heterodera* and *Globodera* induce syncytia consisting of partly fused cells without nuclear division; species of *Meloidogyne* induce giant cells; *Sarisodera* spp. (Heteroderidae) induce a single huge giant cell with a relatively enormous nucleus. This kind of nutritive cell also develops in infections with several other genera of Heteroderidae and Meloidogynidae, as well as with an unrelated nematode, *Rotylenchulus macrodoratus*.

Normal differentiation of xylem and phloem is disturbed in roots infected with *Meloidogyne* spp. Pericycle cells replace some xylem and phloem vessels in galls and root function declines. Because infected roots are undergoing new growth, and transport from root to top is diminished, root to top ratios change in favor of roots. That is, an infected plant devotes more than its usual proportion of energy to building roots at the expense of tops. Such plants have less ability to cope with water stress; top growth and yield are reduced. In addition, photosynthesis proceeds at a lower rate in infected plants and nitrogen- fixing bacteria are inhibited. *Meloidogyne* nematodes often invade nodules and reproduce there, resulting in disturbance of normal function.

Soil conditions have a strong influence on this nematode's reproduction and pathogenicity. In general, lighter soils favor the nematode and clay inhibits it. Thus, in assessing performance of resistant cultivars, tests should be run on various soils with several concentrations of nematode inocula.

Interaction with Other Pathogens

Infection with *Meloidogyne* makes some plants more susceptible to infection with pathogenic fungi. Because contents of exudates from galled roots are altered and quantities are increased, resting stages of fungi within reach may be activated. One thorough study has been made of the mechanism by which *Meloidogyne incognita* alters response of a tomato to the fungus *Rhizoctonia solani* (11). Experiments were based on the observation that both pathogens together induce a severe root rot, but each by itself does not.

Treatment	Root Rot
Rhizoctonia	No
Meloidogyne	No
Rhizoctonia plus *Meloidogyne*	Yes
Rhizoctonia plus leachate from healthy plant	No
Rhizoctonia plus leachate from *Meloidogyne*-infected plant	Yes
Rhizoctonia plus *Meloidogyne* plus continual irrigation to remove leachate	No

By means of radioactive tracers and chemical analysis we know that carbohydrates accumulate in galled roots, especially during the first 2 weeks of infection. Root exudates also contain increased amounts of carbohydrates in comparison with those from uninfected roots. In subsequent weeks, exudates from galled roots contain increased quantities of proteins and amino acids. These changes in organic substances in the soil doubtless influence attraction and growth of fungi and enhance their ability to penetrate roots. A tobacco plant is more susceptible to *Fusarium* when infection with *Meloidogyne* precedes invasion by the fungus (6). Hyphae concentrate in giant cells and galled tissues, kill giant cells, and then invade vascular tissues to grow vigorously into the stem. Giant cells and galls provide food bases from which the fungus rapidly acquires energy to invade other tissues. *Fusarium* behaves similarly in cotton and peas infected with *Meloidogyne*. Cultivars genetically resistant to *Fusarium* succumb to wilt when *Meloidogyne* is present. Not all species of *Meloidogyne* are equally effective, however. *M. hapla* seems less effective than *M. incognita* in breaking genetic resistance of tomato to *Fusarium* wilt. Genetic resistance to infection is also lost in the combination of root knot and *Phytophthora* fungus in black-shank disease of tobacco. In addition, certain fungi not normally pathogenic may become pathogens in the presence of *Meloidogyne*. Seven fungi common in soil, not normally pathogenic to tobacco, induce extensive decay of roots bearing *Meloidogyne* nematodes. This nematode and *Fusarium* together also predispose tobacco to a leaf fungus, *Alternaria* (6).

Interactions between *Meloidogyne* and bacterial pathogens have also been reported. *M. incognita* increases severity of bacterial canker resulting from stem inoculations of tomato with *Corynebacterium michiganense* (2), and *M. javanica* in pot experiments enhances the severity and incidence of gladiolus scab caused by *Pseudomonas marginata* (3).

Control

Early work on nematicides concentrated on root knot because results of treatment were readily seen. Improvement in plant growth and yield after nematicide treatment is often marked. Control of *Meloidogyne* also demonstrates the role of these nematodes in infections with fungi. Nematicides are usually too expensive for crops of low value. Growers on small farms in the tropics lack cash—they cannot use chemical control of root knot. Nematicides are used principally on high-value crops in large agricultural enterprises.

Control measures other than use of nematicides are in great demand. Many breeders throughout the world are seeking genes for resistance to incorporate into crops of their own countries. In addition, careful studies of host range are under way to find suitable crop rotations. Plastic mulch is now being used to heat soil for control of nematodes and other pathogens in countries with abundant sunlight. This technique is called **solarization**. And **biological control**, the employment of pathogens of the nematodes, is under active investigation. Fallow periods, where economically feasible, can depress populations below dangerous levels.

Research

Two approaches to an analysis of nematode secretions that presumably control development and maintenance of giant cells are currently being taken. Hussey and associates are isolating gland contents of infective juveniles by homogenizing large numbers of nematodes and collecting gland contents by differential centrifugation. At the same time they are using histochemical tests for enzyme activity in granules present in gland ducts. They plan to produce monoclonal antibodies to determine whether enzymes are delivered to giant cells and where the enzymes go (8). Veech and associates, however, are incubating adult female *Meloidogyne* in buffered saline and collecting stylet exudates by micromanipulation. They hope to determine enzyme activity in these exudates (5). Both approaches represent application of modern techniques to a central question: How do these nematodes control formation of a feeding site necessary for completion of their life cycle?

Contemporary biotechnology offers the possibility of applying molecular diagnostics to nematodes. *Meloidogyne* species can be differentiated by analysis of mitochondrial and of total DNA. The technique has been more useful for separating species than for separating intraspecific races (1, 7).

The influence of soil type, long known to be important for potential hazard of root knot, is still under study. As new resistant cultivars of various crops become available, it is necessary to determine their performance on existing soil types. Two recent papers illustrate this (4, 13).

An important effort to investigate *Meloidogyne* around the world has been undertaken in recent years. The net result of this will be to accelerate development of resistant crops, especially in the tropics, and to train investigators for crop management (9).

References

1. Curran, J., McClure, M. A., and Webster, J. M. 1986. Genotypic differentiation of *Meloidogyne* populations by detection of restriction fragment length difference in total DNA. J. Nematol. 18:83 – 86.
2. de Moura, R. M., Echandi, E., and Powell, N. T. 1975. Interaction of *Corynebacterium michiganense* and *Meloidogyne incognita* on tomato. Phytopathology 65:1332-1335.

3. El-Goorani, M. A., Abo-El-Dahab, M. K., and Mehiar, F. F. 1974. Interaction between root knot nematode and *Pseudomonas marginata* on gladiolus corms. Phytopathology 64:271 – 272.

4. Niblack, T. L., Hussey, R. S., and Boerma, H. R. 1986. Effects of environment, *Meloidogyne incognita* inoculum levels, and *Glycine max* genotype on root-knot nematode-soybean interactions in field microplots. J. Nematol. 18:338 – 346.

5. Nordgren, R. M., Veech, J. A., and Starr, J. L. 1985. Partial characterization of stylet exudates from *Meloidogyne incognita* and *M. arenaria*. J. Nematol. 17:507 (Abst.)

6. Powell, N. T. 1971. Interactions between nematodes and fungi in disease complexes. Ann. Rev. Phytopathol. 9:253 – 274.

7. Powers, T. O., Platzer, E. G., and Hyman, B. C. 1986. Species-specific restriction site polymorphism in root-knot nematode mitochondrial DNA. J. Nematol. 18:288 – 293.

8. Reddigari, S. R., Sundermann, C. A., and Hussey, R. S. 1985. Isolation of subcellular granules from second-stage juveniles of *Meloidogyne incognita*. J. Nematol. 17:482 – 488.

9. Sasser, J. N., Eisenback, J. D., Carter, C. C., and Triantaphyllou, A. C. 1983. The international *Meloidogyne* project -- its goals and accomplishments. Ann. Rev. Phytopathol. 21:271 – 288.

10. Van Gundy, S. D., Bird, A. F., and Wallace, H. R. 1967. Aging and starvation in larvae of *Meloidogyne javanica* and *Tylenchulus semipenetrans*. Phytopathology 67:559 – 571.

11. Van Gundy, S. D., Kirkpatrick, J. D., & Golden, J. 1977. The nature and role of metabolic leakage from root-knot nematode galls and infection by *Rhizoctonia solani*. J. Nematol. 9:113 – 121.

12. Williams, T. D. and Siddiqi, M. R. 1972. Commonwealth Institute of Helminthology, Descriptions of Plant-parasitic Nematodes, Set 1, No. 3: *Meloidogyne javanica*; see also Set 3, No. 31: *M. hapla*; Set 4, No. 49: *M. exigua*; Set 5, No. 62: *M. arenaria*; Set 6, No. 87: *M. graminicola*.

13. Windham, G. L. and Barker, K. R. 1986. Effects of soil type on the damage potential of *Meloidogyne incognita* on soybean. J. Nematol. 18:331 – 338.

CRICONEMOIDES AND RELATED GENERA

Introduction

Individuals of these genera are called "ring nematodes" in reference to broad annulations appearing like rings surrounding the body. They are slow-moving ectoparasites with long, stout stylets. High populations can severely damage roots. Various species reproduce on grape, pine, elm, maple, and peach. In addition, some species are ectoparasites of strawberry, carnation, grasses, Chinese cabbage, and eggplant.

Classification

Order Tylenchida; suborder Tylenchina; superfamily Criconematoidea; family Criconematidae; subfamily Criconematinae.

Morphology

Nematodes of this group are short and thick with conspicuous broad annules; sexes are separate in some species but males are unknown in others. Females of *Macroposthonia xenoplax* (formerly *Criconemoides xenoplax*) are 400 – 600 μm long, with a length/width ratio of 8 – 13.6. Their bodies are cylindrical, tapering only slightly at head and tail. The head is broad, with elevated lips, but not offset. The stylet is stout and long (71 – 86 μm) with prominent knobs. Members of the family Criconematidae can be recognized by the fused procorpus-metacorpus. The procorpus merges directly into a large metacorpus with conspicuous crescentic plates. Esophageal glands are in a small basal bulb. The vulva is posterior, at 90 – 95% of the body length. There is a single gonad. Broad annules with smooth or ornamented posterior margins are separated by deep divisions. The short tail is rounded. Adult males are degenerate, often lacking a stylet. They do not have caudal alae (3). See Fig. 6.30 for the diagnostic characters.

The superfamily Criconematoidea includes nematodes with an esophagus of the type described above. Three families of Criconematoidea are readily distinguished:

Family Criconematidae: with prominent annulations and large thick stylets.

Family Paratylenchidae: with fine annulations and elongate, fine stylets.

Family Tylenchulidae: with very short stylets and swollen females.

In the family Criconematidae the subfamily Criconematinae includes a number of genera, whose names have changed as the taxonomy continues to be revised. The list includes *Criconema, Criconemella, Macroposthonia*, and *Criconemoides*. Useful references for details of morphology, including keys to species are (4), (5), (7), and (8).

Biology

Species of the subfamily Criconematinae are ectoparasites with an unusual method of locomotion. Instead of the typical undulations, these nematodes generate waves of elongation and contraction that pass from the tail forward along the body. The coarse annules presumably provide traction against soil particles. Extraction from soil by the Baermann funnel is usually unsuccessful. Centrifugal flotation is more useful.

These ectoparasites feed on outermost root tissues and reach dense populations around roots of fruit trees and other woody plants. They re-

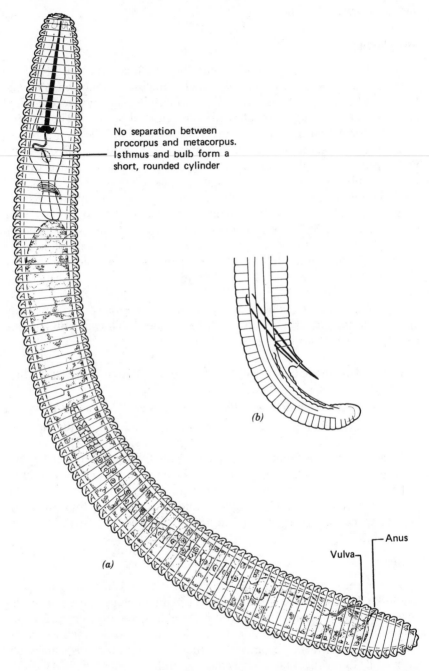

No separation between procorpus and metacorpus. Isthmus and bulb form a short, rounded cylinder

(b)

(a)

Anus

Vulva

Fig. 6.30 *Macroposthonia xenoplax.* (**a**) Adult female (× 388). Note long, thick stylet; body width relatively great in proportion to length; annulations widely spaced. (**b**) Male posterior (×790). (From K. J. Orton Williams, 1972. *Macroposthonia xenoplax.* Commonwealth Institute of Helminthology, Descriptions of Plant-parasitic Nematodes, Set 1, No. 12, © Commonwealth Agricultural Bureaux, by permission.)

main attached to the root epidermis while the strong stylet penetrates cells of the outer root tissues. Anterior portions of some specimens may penetrate roots, or occasionally the entire nematode may enter. The nematodes are usually found in sandy soils where moisture is maintained.

Pathology

Necrotic lesions result from death of depleted cells, and other organisms add to the damage. Destruction of new growth of roots affects a plant's total physiology, resulting in increased moisture stress, reduced free amino acid contents of roots and shoots, changes in the relative proportions of individual amino acids, and alterations in glucoside concentrations (6). Criconematid nematodes have been suspected as transmitters of plant viruses, although they are not the major vectors (*Xiphinema* spp., *Longidorus* spp. and *Trichodorus* spp.).

Interaction with Other Pathogens

Species of *Macroposthonia* (= *Criconemoides*) are suspected of contributing to "peach tree short life." In certain orchards, peach trees begin growth normally, then die in a few years. In others, trees continue bearing for many years. The problem seems to involve frost damage, attack by several root pathogens, poor soil, and criconematids. By inhibiting normal root growth and function, the ectoparasites stress plants and make them more vulnerable to other stresses. Increased susceptibility of plum trees to bacterial and *Cytospora* cankers as well as increased damage to pecans from *Pythium* and *Fusarium* fungi are associated with nematode infection.

Control

Nematicides depress populations of Criconematidae in peach orchards. Tree vigor improves and losses from short-life disease decline. Because many of these nematodes have wide host ranges, weed control is essential for successful nematode control in rice (2), and undoubtedly in other crops. Some genetic resistance is incorporated into cultivars of pearl millet, maize, and soybeans, but much remains to be done.

Research

Practical field problems are related to the physiological characteristics of nematode species. *Belonolaimus longicaudatus*, a destructive pest, reproduces at soil temperatures up to 34°C, but is inhibited at higher temperatures. *Criconemoides* sp., a relatively mild pest, is not hindered up to 41°C (1). In the presence of both nematodes, certain pasture grasses yield best at higher temperatures.

References

1. Boyd, F. T., Schroder, V. N., and Perry, V. G. 1972. Interaction of nematodes and soil temperature on growth of three tropical grasses. Agronomy J. 64:497 – 500.
2. Hollis, J. P. 1972. Nematocide—weeds interaction in rice fields. Plant Dis. Rep. 56: 420 – 424.
3. Loof, P. A. A. 1974. Commonwealth Institute of Helminthology Descriptions of Plant-parasitic Nematodes Set 3, No. 42: *Criconemoides morgensis*; see also Set 1, No. 12: *Macroposthonia xenoplax*; Set 2, No. 28: *M. sphaerocephala*; Set 4, No. 57: *Criconema palmatum*; Set 4, No. 58: *M. curvata.*
4. Loof, P. A. A. and De Grisse, A. 1972. Interrelationships of the genera of Criconematidae (Nematoda: Tylenchida). Meded. Fak. Landbouwwet. Rijksuniv. Gent 38:1303 – 1328.
5. Luc, M. 1970. Contribution a l'étude du genre Criconemoides Taylor, 1936 (Nematoda: Criconematidae). Cah. ORSTOM, Ser. Biol. 11:69 – 150.
6. Okie, W. R. and Reilly, C. C. 1984. Effect of the ring nematode upon growth and physiology of peach rootstocks under greenhouse conditions. Phytopathology 74:1304 – 1307.
7. Raski, D. J. and Golden, A. M. 1965. Studies on the genus *Criconemoides* Taylor, 1936 with descriptions of eleven new species and *Bakernema variable* n.sp. (Criconematidae: Nematoda). Nematologica 11:501 – 565.
8. Taylor, A. L. 1936. The genera and species of the Criconematinae, a sub-family of the Anguillulinidae (Nematoda). Trans. A. Microsc. Soc. 55:391 – 421.

HEMICYCLIOPHORA

Introduction

Hemicycliophora spp. are widely distributed and damage certain citrus spp., some grapes, melons, carrots, and pine.

Classification

Order Tylenchida; suborder Tylenchina; superfamily Criconematoidea; family Criconematidae; subfamily Hemicycliophorinae.

Morphology

These elongate nematodes are readily recognized by their coarsely annulated double cuticle. Sexes are separate, and differ greatly in appearance. Females are 0.4—2 mm long, with a length/width ratio of approximately 25. Lip region is bluntly rounded, not offset. The long stylet has a tapering anterior section longer than the knobbed posterior portion. Knobs slope posteriorly. Isthmus of the esophagus is broad and short. The excretory pore is at the level of the esophago-intestinal junction. The vulva has protruding lips and is located posteriorly at 75—90% of the body length. A single gonad extends anteriorly and is not reflexed. The tail varies among species and is the region where the cuticular sheath is most conspicuous. In *Hemicycliophora arenaria* the tail constitutes from one-fourth to one-seventh the body length. *Hemicriconemoides* is a related genus with a double cuticle.

Males are thinner than females, lack a stylet, and have a degenerate esophagus. Under the light microscope they do not appear to have a cuticular sheath; but the electron microscope reveals its presence. The testis is single and spicules are sickle-shaped. The tail is long and slender, and the bursa is confined to the region of the cloaca (2). A key to species appears in Ref. (1). See Fig. 6.31 for morphology of this genus.

Biology

Hemicycliophora spp. occur in most regions of the world, generally in sandy soils and on a wide variety of hosts. *H. arenaria* feeds as an ectoparasite on root tips and is attached during feeding by a polysaccharide plug at its lips. The outermost cuticle is a thin sheath secreted along with the cuticle at each molt (3, 4).

Pathology

The stylet enters cortical cells at tips of tap and lateral roots, and suction developed by the median bulb withdraws the cytoplasm. Cells at feeding sites are killed. Details of this process are described in Ref. (5). *H. arenaria* stimulates pericycle tissue to initiate new growth; root tips are galled. Figure 6.32 compares galls induced by a species of this genus with root-knot galls. Hosts show the expected poor growth and yield resulting from impaired root function.

Interaction with Other Pathogens

None reported.

Control

Hemicycliophora nematodes are usually found protruding from roots; nematicides are effective. In addition, woody plants can be dipped in hot water (for citrus nursery stock, 10 min at 46°C is effective).

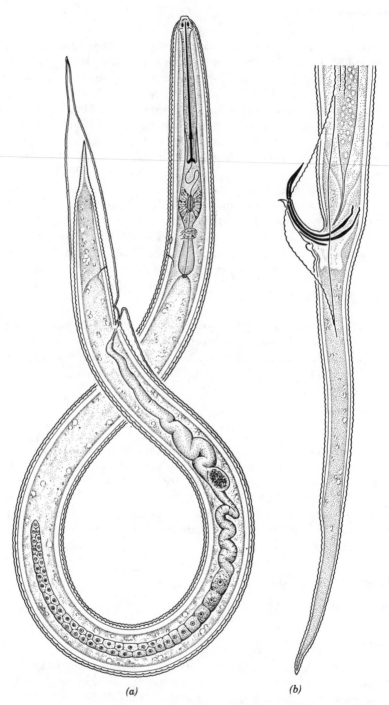

Fig. 6.31 *Hemicycliophora vitiensis* (**a**) Female (×410). (**b**) Male, posterior end (×840). (From K. J. Orton Williams. 1978. Two new species of the genus *Hemicycliophora* De Man, 1921 (Nematoda: Tylenchida). Rev. Nématol. 1:197 – 205, by permission.)

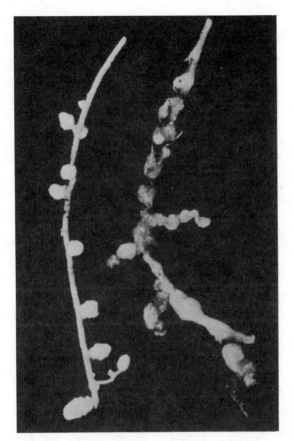

Fig. 6.32 Comparison of galls caused by *Hemicycliophora arenaria* on root tips of rough lemon (left) with those caused by *Meloidogyne javanica* on tomato (right). (From M. T. Franklin, and A. R. Stone, 1974. *Hemicycliophora arenaria.* Commonwealth Institute of Helminthology, Descriptions of Plant-parasitic Nematodes, Set 3, No. 43, © Commonwealth Agricultural Bureaux, by permission.)

Research

H. arenaria is more sensitive to low oxygen diffusion rates in soil than some other phytonematodes. During irrigation and for a few days subsequently, oxygen levels are sufficiently low to inhibit reproduction. Populations in citrus orchards may be kept in check by frequent irrigation (6 – 8). Recent publications have dealt with taxonomy and host range in Africa, India, Australia, New Zealand, and Europe.

References

1. Brzesky, M. W. and Ivanova, T. S. 1978. Taxonomic notes on *Hemicycliophora* de Man (Nematoda: Hemicycliophoridae). Nematol. Medit. 6:147 – 162.

2. Franklin, M. T. and Stone, A. R. 1974. *Hemicycliophora arenaria*. Commonwealth Institute of Helminthology Descriptions of Plant-parasitic Nematodes, Set 3, No. 43.

3. Johnson, P. W., Van Gundy, S. D., and Thomson, W. W. 1970. Cuticle ultrastructure of *Hemicycliophora arenaria, Aphelenchus avenae, Hirschmaniella gracilis* and *Hirschmaniella belli*. J. Nematol. 2:42 – 58.

4. Johnson, P. W., Van Gundy, S. D., and Thomson, W. W. 1970. Cuticle formation in *Hemicycliophoria arenaria, Aphelenchus avenae* and *Hirschmaniella belli*. J. Nematol. 2: 59 – 79.

5. McElroy, F. D. and Van Gundy, S. D. 1968. Observations on the feeding processes of *Hemicycliophora arenaria*. Phytopathology 58:1558 – 1565.

6. Van Gundy, S. D., Rackham, R. L. 1961. Studies on the biology and pathogenicity of *Hemicycliophora arenaria*. Phytopathology 51:393 – 397.

7. Van Gundy, S. D. and Stolzy, L. H. 1963. The relationship of oxygen diffusion rates to the survival, movement, and reproduction of *Hemicycliophora arenaria*. Nematologica 9:605 – 612.

8. Van Gundy, S. D., McElroy, F. D., Cooper, A. F., and Stolzy, L. H. 1968. Influence of soil temperature, irrigation and aeration on *Hemicycliophora arenaria*. Soil Sci. 106: 270 – 274.

PARATYLENCHUS

Introduction

Species of this genus are small and have long, delicate stylets. Although each individual is relatively harmless, large populations are sometimes dangerous to crop production. Hosts include plants in the families Cruciferae and Umbelliferae, such as carrot, celeriac, parsley, and cabbage. Some forage and woody plants, including clover, grasses, grapes, and fruit trees, are also hosts.

Classification

Order Tylenchida; suborder Tylenchina; superfamily Criconematoidea; family Paratylenchidae; subfamily Paratylenchinae.

Morphology

These are the smallest phytonematodes with adults usually less than 0.5 mm long. Sexes are separate. *Paratylenchus projectus* females measure from 300 to 422 μm long with a ratio of total length to width of 21.5 – 28.

Lip region is offset and shaped like a truncated cone. The stylet is fine, with rounded knobs, and measures about 25 μm. The narrow procorpus of the esophagus merges into a large elongate metacorpus with prominent crescentic plates. The isthmus is long and narrow, and esophageal glands are in a small bulb. The vulva is located about 83% of the body length and the gonad is single. Cuticular annulations are narrow, with delicate divisions. There are four lateral lines in many species. The tail curves ventrally and equals from one-sixth to one-eighth of the body length. Adults and preadults are recognizable at low magnifications by their size and by a characteristic, ventrally curved position. In some species the body narrows just behind the vulva. Males are degenerate, lacking a stylet. The tail has no caudal alae (5). Figure 6.33 illustrates the morphology of this genus.

In addition to the genus *Paratylenchus*, this subfamily includes *Gracilacus*, with species having very long stylets, and *Cacopaurus*, whose females are obese and sedentary.

Biology

Second- and third-stage juveniles, as well as adult females of *Paratylenchus*, feed. However, J-4s lack stylets and survive in soil for long periods without feeding. This stage requires root exudates from hosts to make its final molt. Young adults enter roots and large populations often develop, especially in fine-textured soil.

Pathology

Paratylenchus nematodes insert their stylets into epidermal cells or the base of a root hair. Individuals may feed on the same cell for several days without killing it. However in certain hosts, feeding by *Paratylenchus* spp. results in brown necrotic areas, and massive populations can kill roots.

Interaction with Other Pathogens

There are a few examples of enhanced damage from fungi in the presence of high populations of *Paratylenchus* spp. Examples of this interaction are: *Lophodermium pinaster* on *Pinus sylvestris* seedlings, and *Phoma* spp. on mint.

Control

Varietal differences in susceptibility are known and genetic control seems feasible, although plant breeders have not yet devoted major efforts to

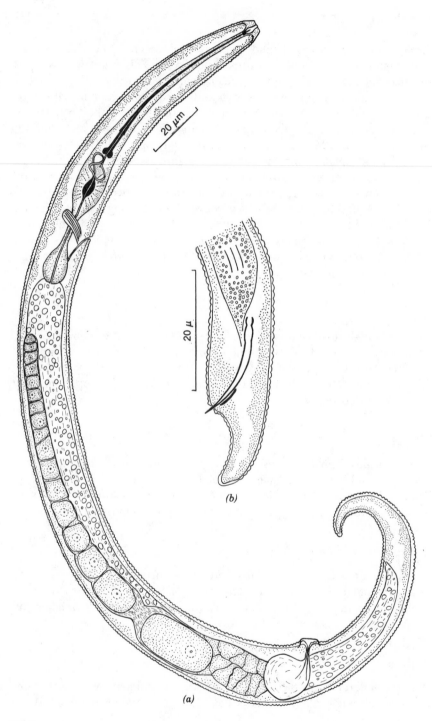

(a)

(b)

20 μm

20 μ

obtain resistance to *Paratylenchus*. Many field trials of nematicides of several types have been successful. Crop rotation is difficult, because arrested J-4s may sruvive starvation for more than a year.

Research

Feeding behavior has attracted attention. The long stylet of a female *Gracilacus* sp. attaches the nematode to olive roots and is enveloped by a feeding tube in the cell (3). Similar feeding tubes are formed at the surface of cells penetrated by *Trichodorus*. Such tubes are probably formed from coagulated secretions from amphids.

Many phytonematodes are distributed to irrigated fields in water that has drained from cropland. In a field plot *Paratylenchus* sp. reached soil that had never received irrigation water. This nematode was probably carried by wind (2). Populations decline in soils treated with nematicides only to recur in greater numbers afterward.

Arrested development at the preadult stage is an example of the same kind of requirement for a signal from growing roots found in *Heterodera*. Nature of the stimulus to *Paratylenchus* is unknown (4).

Species of these tiny nematodes appear to be more resistant to certain nematicides than other phytonematodes. Vorlex at 112 L/ha controlled *Pratylenchus penetrans* in soil prepared for fruit trees. but twice that dose failed to affect *Paratylenchus* sp. (6). Similar results were reported with methyl bromide, a general biocide (1).

Taxonomy of the genus has received much attention, especially from Raski (7 – 9).

References

1. Abdallah, N. and Lear, B. 1975. Lethal dosages of methyl bromide for four plant-parasitic nematodes and the effect of soil temperature on its nematicidal activity. Pl. Dis. Rep. 59:224 – 228.

2. Faulkner, L. R. and Bolander, W. J. 1970. Agriculturally-polluted irrigation water as a source of plant-parasitic nematode infestation. J. Nematol. 2:368 – 374.

3. Inserra, R. N. and Vovlas, N. 1977. Parasitic habits of *Gracilacus peratica* on olive feeder roots. Nematol. Mediterr. 5:345 – 348.

Fig. 6.33 *Paratylenchus marylandicus.* (a) Adult female (×800); (b) Male tail (×1600). Note elongate stylet of female; fine annulations; body coiled or in form of the letter C after immobilization by mild heat; fused procorpus and metacorpus. (From W. R. Jenkins. 1973. *Paratylenchus marylandicus,* n. sp. (Nematoda: Criconematidae) associated with roots of pine. Nematologica 5:175 – 177, by permission of E. J. Brill.)

4. Ishibashi, N., Kondo, E., and Kashio, T. 1975. The induced molting of 4th-stage larvae of pin nematode, *Paratylenchus aciculus* Brown (Nematoda: Paratylenchidae), by root exudate of host plant. Appl. Entomol. Zool. 10:275 – 283.

5. Loof, P. A. A. 1975. Commonwealth Institute of Helminthology Descriptions of Plant-parasitic Nematodes, Set 5, No. 71: *Paratylenchus projectus*; see also Set 6, No. 79: *P. bukowinensis.*

6. Marks, C. F. and Davidson, T. R. 1973. Effects of preplant and postplant nematicides on populations of nematodes in the soil and on growth of fruit trees in the Niagara Peninsula. Canad. Plant Dis. Surv. 53:170 – 174.

7. Raski, D. J. 1975. Revision of the genus *Paratylenchus* Micoletzky, 1922 and descriptions of new species. Part I of 3 parts. J. Nematol. 7:15 – 34.

8. Raski, D. J. 1975. Revision of the genus *Paratylenchus* Micoletzky, 1922 and descriptions of new species. Part II of three parts. Jour. of Nematol. 7:274 – 295.

9. Raski, D. J. 1976. Revision of the genus *Paratylenchus* Micoletzky, 1922 and descriptions of new species. Part III of three parts—*Gracilacus*. J. Nematol. 8:97 – 115.

TYLENCHULUS

Introduction

Tylenchulus semipenetrans, a sedentary nematode, parasitizes citrus in most areas where this crop is grown. In addition to citrus, *T. semipenetrans* reproduces on olive and grape. It is an important pest of grapes in Australia.

Classification

Order Tylenchida; suborder Tylenchina; superfamily Criconematoidea; family Tylenchulidae; subfamily Tylenchulinae.

Morphology

This species is small; adult females measure about 375 µm long, with a length/width ratio of 4.5. Adult females have a long narrow anterior portion and a swollen posterior (Fig. 6.34). The head is not offset in either sex. Stylet is 13 µm long, with well developed rounded basal knobs; esophageal glands are in a bulb, abutting the intestine. There is a single, coiled ovary. The vulva has thick lips and is located near the posterior end. The excretory pore is in an unusual position, just anterior to the vulva. The intestine lacks a lumen; rectum and anus are absent. Males are about 370 µm long, with a length/width ratio of about 40. Stylet and esophagus are degenerate (4).

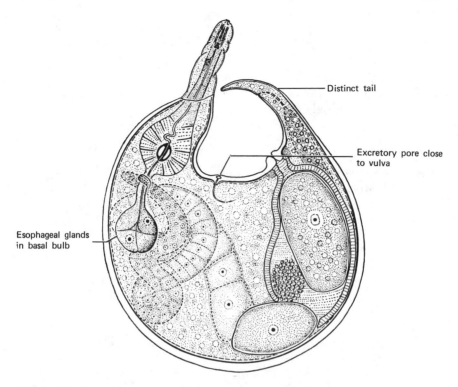

Distinct tail

Excretory pore close
to vulva

Esophageal glands
in basal bulb

Fig. 6.34 *Tylenchulus mangenoti,* adult female, parasite of a tropical herb (×570). (From M. Luc. 1957. *Tylenchulus mangenoti* n. sp. (Nematoda-Tylenchulidae). Nematologica 2:329 – 334, by permission of E. J. Brill.)

Biology

Female juveniles feed on root surface cells. Adult females penetrate part way into roots, leaving their posteriors outside. The elongate neck swells irregularly as it grows and·is subjected to pressures from root cortex cells. The external posterior portion swells and becomes covered with a gelatinous material extruded from the excretory pore. Eggs are deposited into this mucoid mass. Male juveniles and adults do not feed. Reproduction may be either with or without males.

There are populations of this nematode around citrus plants of all kinds, wherever citrus is grown. At each spring flush of root growth, the numbers of nematodes increase. *T. semipenetrans* occurs in soil wherever host roots are present. They have been recovered from depths of up to 4 m, but the greatest concentrations are within the upper 30 cm of soil. Second-stage juveniles may arrest their development and persist in soil without host plants for several years. When host roots become available the nematodes proceed to maturity (Fig. 6.35).

Within the species *T. semipenetrans*, several races with different host ranges are present. Six different populations have been identified, one of which reproduces only on a grass.

Pathology

Unthrifty growth of citrus resulting from infection with this nematode is called "slow decline" in contrast with "spreading decline," the more drastic effect of attack by *Radopholus similis* (cf. Table 6.1). Trees showing slow decline have yellow leaves, twig dieback, small fruits, and do not produce

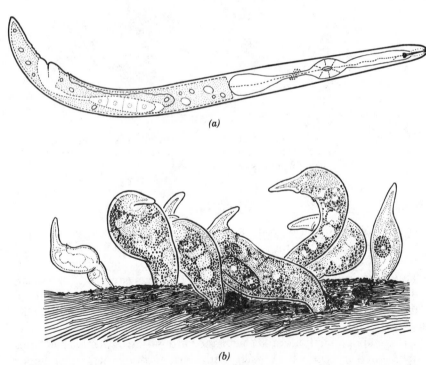

(a)

(b)

Fig. 6.35 *Tylenchulus semipenetrans.* (a) Young adult female (×460). (From S. D. Van Gundy. 1958. The life history of the citrus nematode *Tylenchulus semipenetrans* Cobb. Nematologica 3:283 – 294, by permission of E. J. Brill.); (b) Females on surface of citrus root. (From G. Thorne. 1961. Principles of Nematology. McGraw-Hill, New York, 553 pp., by permission.)

vigorous new growth in the spring. Numbers of nematodes reach very high levels, up to 100 nematodes per centimeter of root. A young female penetrates the cortex and destroys one cell, leaving a space. Six to ten cells surrounding the head are transformed into "nurse cells." Walls thicken, vacuoles are lost, cytoplasm becomes dense, and nuclei enlarge. The nematode feeds on these transformed cells. No galls develop and nurse cells retain their normal size. Eventually they break down and microorganisms invade the tissue. Because a citrus grove is expected to produce good yields for many years, risk of infection with *T. semipenentrans* and resulting yield loss are of great concern to growers. Figure 6.36 is a three-dimensional diagram of *T. semipenetrans* in a citrus root.

Interaction with Other Pathogens

Fusarium oxysporum and *F. solani* increase damage from *T. semipenetrans*.

Control

Strict sanitation must be practiced in the movement of nursery stock to establish or replant orchards. Chemical control has been successful, especially by incorporation of nematicide in irrigation water. Citrus breeders have produced resistant rootstocks on which desirable citrus scions are grafted.

Research

Researchers in many parts of the world continue to test various nematicides for control of nematodes on citrus. This has become important in recent years in the United States because one of the best compounds, DBCP, was removed from commerce because of its hazard to health. Nitrogen fertilizers may be useful as alternatives to nematicides. In one study, application of ammonium nitrate, ammonium sulfate, or urea to infected citrus reduced the numbers of nematodes and improved plant growth. The authors attributed this effect to an increase in the synthesis of phenols that are antagonistic to the nematodes (1).

Mycorrhizal fungi have symbiotic associations with many trees. These fungi enhance absorptive powers of roots, especially for phosphorus, and thereby stimulate growth of the host. One of the harmful effects of nematicides is that they may kill the fungi and harm the plant. *Glomus mossae* improved growth of lemon seedlings, whether they were infected by nematodes or not (3).

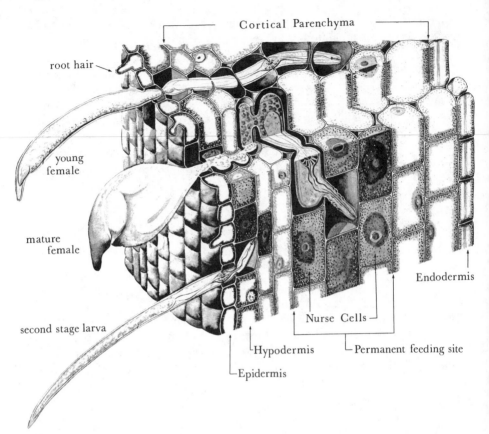

Fig. 6.36 *Tylenchulus semipenetrans.* Diagrammatic drawing illustrating the general cellular arrangement of the root tissue involved in citrus nematode parasitism; cell response to nematode feeding; and the three phases of parasitism by the citrus nematode: juvenile feeding, penetration into the inner cortex by the young female by the stretching of her anterior body portion, and the established female in permanent feeding and egg laying position. Based on an average body length, it would take 80 second-stage juveniles laid head to tail to measure 1 inch. (From S. D. Van Gundy and J. D. Kirkpatrick. 1965. Factors explaining citrus nematode resistance. California Citrograph 50:235 – 241. by permission.)

Four cellular responses to infection by *T. semipenetrans* appear in resistant citrus. One or more of these may occur in a particular resistant host.

(a) Cells around the nematode die rapidly in resistant responses to other nematodes as well as to *T. semipenetrans.*

(b) Wound periderm forms in some resistant associations. This tissue is believed to participate in producing antinematode compounds.

(c) In some nematode - plant combinations, cavities develop around the invaders. There are associated with rapid necrosis of cells in one host but occur without rapid cell necrosis in another.

(d) Nurse cells in susceptible citrus have dense cytoplasm without conspicuous vacuoles. But in a resistant combination, nurse cells have large and numerous vacuoles and the nematodes do not reproduce. Probably the presence of vacuoles indicates failure of nurse cells to respond to nematode demands for nutrition (2).

A prime question in current plant pathology is: How do resistant hosts recognize certain invading pathogens as foreigners whereas susceptible hosts fail to do so? Sugar residues on the surface of some phytonematodes, including *T. semipenetrans*, may be important in this process. Investigators identified sialic acid and galactose/N-acetylgalactosamine residues on the nematode cuticles by using antibodies carrying a fluorescent dye. This work suggests a way of describing surface properties of nematodes that may play a role in host recognition (5).

References

1. Badra, T. and Elgindi, D. M. 1979. The relationship between phenolic content and *Tylenchulus semipenetrans* populations in nitrogen-amended citrus plants. Rev. Nématol. 2:161 – 164.
2. Kaplan, D. T. 1981. Characterization of Citrus rootstock responses to *Tylenchulus semipenetrans* (Cobb). J. Nematol. 13:492 – 498.
3. O'Bannon, J. H., Inserra, R. N., Nemec, S., and Vovlas, N. 1979. The influence of *Glomus mossae* on *Tylenchulus semipenetrans*-infected and uninfected *Citrus limon* seedlings. J. Nematol. 11:247 – 250.
4. Siddiqi, M. R. 1974. Commonwealth Institute of Helminthology Description of Plant-parasitic Nematodes, Set 3, No. 34: *Tylenchulus semipenetrans*.
5. Spiegel, Y., Cohn, E., and Spiegel, S. 1982. Characterization of Sialyl and Galactosyl residues on the body wall of different plant parasitic nematodes. J. Nematol. 14:33 – 39.

APHELENCHOIDES

Introduction

The genus *Aphelenchoides* includes some nematodes living in soil and others on aerial parts of plants. The species have a great variety of feeding habits: some feed on developing leaves and flowers in buds and others enter stomata to live within leaves. A few species cause important problems on strawberries, rice, ferns, and ornamentals.

Classification

Order Tylenchida; suborder Aphelenchina; superfamily Aphelenchoidea; family Aphelenchoididae.

Morphology

Sexes are separate in many species but males are rare in others. Lengths of females range from 0.4 to 1.2 mm with a length/width ratio from 32 – 65. The head is flattened or rounded anteriorly and slightly offset from the body. Stylets vary from about 10 to 17 μm long with needlelike anterior portions and small to minute knobs posteriorly. The esophagus has a prominent median bulb; glands overlap the intestine dorsally in a long lobe. Duct openings are in the median bulb so that the esophageal lumen behind the stylet does not have a Y-connection to the gland duct characteristic of the superfamily Tylenchoidea. The excretory pore opens at the level of the nerve ring. The vulva is located approximately two- thirds of the body length from the anterior end and there is a single outstretched gonad that may have a reflexed ovary. A long uterine sac extends behind the vulva. The tail is conoid, and its length is up to five times the body width at the anus. It may end in a sharp tip (mucro), with one or more pointed processes, or the tail may have a simple, blunt spike. The cuticle has fine annulations and the lateral field has up to four incisures. Males are very similar in general size and form except for secondary sex characters. Spicules are shaped like a rose thorn (2). The morphology of this genus is shown in Fig. 6.37.

Biology

Many species feed on fungi, and some are both plant parasitic and mycophagous. Nematodes of this genus feed by puncturing cells and withdrawing their contents. *A. ritzemabosi* enters leaves of chrysanthemums through stomata to feed on mesophyll cells. It also invades buds and growing points. This species resists desiccation very well and survives winter in dried leaves. When the temperature rises, the nematodes ascend films of water on stems to enter growing leaves or buds. *A. besseyi* is ectoparasitic on strawberry buds and leaves and is also an important pest of rice. It survives between crops in a dormant condition under hulls of rice grains. When films of moisture are present, nematodes migrate to growing points of leaves and stems. The life cycle is shown in Fig. 6.38.

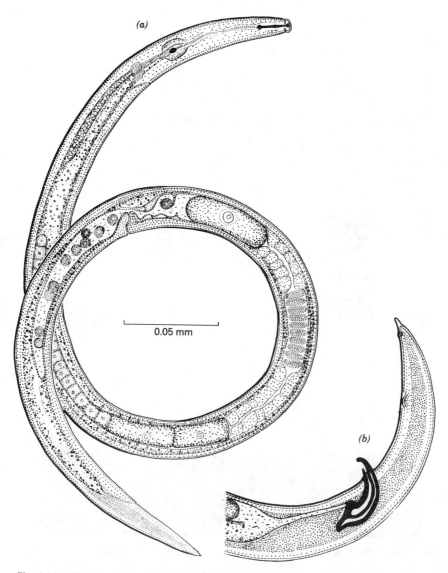

Fig. 6.37 *Aphelenchoides blastophthorus* (**a**) Adult female (×520). Note the delicate stylet, small knobs, esophageal lumen without bend behind stylet, large metacorpus joining intestine directly; esophageal glands overlapping intestine in a long lobe; one outstretched ovary, long postvulval sac. (**b**) Male tail, lateral (×855). (From D. J. Hooper. 1975. *Aphelenchoides blastophthorus*, Commonwealth Institute of Helminthology, Set 5, No.73, © Commonwealth Agricultural Bureaux, by permission.)

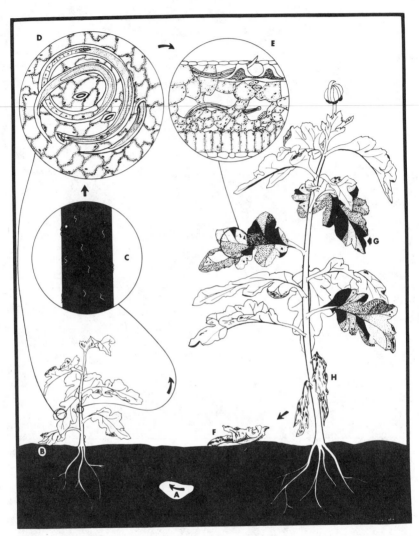

Fig. 6.38 *Aphelenchoides ritzemabosi*—life cycle on chrysanthemum. (**a**) Nematodes move over soil surface to chrysanthemum plant (not shown); (**b**) and (**c**) ascend stem; (**d**) enter through stomata; (**e**) invade leaf mesophyll; (**f**) Leaf falls to soil surface; (**g**) Sectors of necrotic tissue; (**h**) Desiccated leaf. (From R. P. Esser. 1966. Life History of a Foliar Nematode (*Aphelenchoides ritzema-bosi*) on Chrysanthemum. Nematol. Circular No. 7. Div. Plant Ind. Florida Dept. of Agriculture and Consumer Services, Gainesville, by permission.)

Pathology

Nematode destruction of cells in buds and growing points of stems disrupts normal development of host plants. Consequently, leaves are malformed, stems are stunted, flower production is reduced, and yield losses ensue. In addition, affected rice plants lose chlorophyll at leaf extremities ("white tip"). Endoparasitic feeding, as in chrysanthemum leaves, results in necrotic, desiccated areas usually delimited by veins. Affected plants have darkened leaf sectors where the nematodes have destroyed mesophyll tissues.

Interaction with Other Pathogens

Plants harboring *Aphelenchoides* spp. suffer increased damage from bacteria and fungi. In at least one association, nematodes protect plants from harmful fungi. Infection of begonias with *A. fragariae* plus *Xanthomonas begoniae* causes rapid, severe leaf necrosis in 10 – 14 days. Without bacteria, nematodes alone induce leaf reddening. Bacteria alone induce small, water-soaked lesions at leaf margins. After 3 – 4 weeks, the entire leaf blade becomes dry and brown (4). Strawberry plants attacked by nematodes alone do not show severe symptoms but have abnormally narrow leaves. Bacteria alone have little effect. Plants infected with both *A. fragariae* and certain strains of *Corynebacterium fascians* resemble small cauliflowers. In extreme cases, plants are stunted and flowers are deformed. Petals fail to form or become small and greenish. Sepals are enlarged and stamens and receptacle are malformed so that strawberries do not develop. Axillary buds are continually produced in the crowns (1).

Control

Dissemination of *Aphelenchoides* spp. can be prevented by sanitation during propagation. Hot water treatment (47°C for 15 min. or shorter periods at higher temperatures) kills nematodes without harming dormant strawberry plants. Foliar sprays (parathion and others) are also effective, as are systemic chemicals (e.g.,oxamyl). White tip of rice, formerly an important problem, is now controlled by warm water or chemical treatment of seed. Moreover, resistant rice and strawberry cultivars have been developed.

Research

In addition to interactions between *Aphelenchoides* and bacteria, there are examples of interactions between these nematodes and fungi. In a susceptible rice cultivar, reduction of plant weight by *Sclerotium oryzae* alone is moderate; and by the nematode alone, also moderate. Both pathogens together cause severe weight reduction. Nematodes reproduce much better on rice plants with the fungus than without. The nematodes probably feed on both fungus and plant (3). Fungus-feeding *Aphelenchoides* protect barley and wheat from footrot caused by *Fusarium culmorum* and *Gerlachia nivalis* (5).

References

1. Crosse, J. E. and Pitcher, R. S. 1952. Studies in the relationship of eelworms and bacteria to certain plant diseases. I. The etiology of strawberry cauliflower disease. Ann. Appl. Biol. 39:475 – 484.

2. Franklin, M. T. and Siddiqi, M. R. 1972. Commonwealth Institute of Helminthology, Descriptions of Plant-parasitic Nematodes, Set 1, No. 4: *Aphelenchoides besseyi*; see also Set 5, No. 73: *A. blastophthorus*; Set 5, No. 74: *A. fragariae*; and Set 6, No. 84: *A. bicaudatus*.

3. McGawley, E. C., Rush, M. C., and Hollis, J. P. 1984. Occurrence of *Aphelenchoides besseyi* in Louisiana rice seed and its interaction with *Sclerotium oryzae* in selected cultivars. J. Nematol. 16:65 – 68.

4. Riedel, R. M. and Larsen, P. O. 1974. Interrelationship of *Aphelenchoides fragariae* and *Xanthomonas begoniae* on Rieger begonia. J. Nematol. 6:215 – 216.

5. Rössner, J. and Urland, K. 1983. Mycophagous nematodes of the genus *Aphelenchoides* from the stem base of cereal plants and their action against foot rot diseases of cereals. Nematologica 29:454 – 462.

BURSAPHELENCHUS

Introduction

The genus *Bursaphelenchus* is associated with bark beetles and with long-horned beetles of conifers. Most species feed on fungi growing in insect tunnels and are carried externally on the beetles. One species, the pine wilt nematode, kills pine trees and has recently destroyed large portions of pine forests in Japan.

Classification

Order Tylenchida; suborder Aphelenchina; superfamily Aphelenchoidea; family Aphelenchoididae.

Morphology

This account applies to one species, *B. xylophilus*, a pathogen of coniferous trees. Sexes are separate. Females average 0.81 mm in length with a length/

width ratio of about 40. Lips are high, rounded, and the head is offset. The stylet is 15 – 16 μm long, with small basal knobs. A large median bulb fills almost the entire width of the body. Esophageal glands are slender, overlapping the intestine dorsally. Excretory pore is at the level of the esophago-intestinal junction. The vulva is at about three-fourths the distance from the anterior. It has a wide, overlapping anterior lip. The single gonad is outstretched, not reflexed. A long postuterine sac extends three-fourths the distance to the anus. The tail is broadly rounded with or without a fingerlike projection. Males are somewhat shorter than females and resemble them except for secondary sex characters. Spicules have a distinctive broad proximal portion and a slightly flared terminal portion. The tail is pointed in the male and curved ventrally. It has small, oval, terminal bursae (8). This description is illustrated in Fig. 6.39.

Biology

All species in this genus are associated with beetles. The life cycle of *B. xylophilus* alternates between pine trees and long-horned, wood-boring beetles (family Cerambycidae). Adult beetles emerge in late spring from dead nematode-infested pines; the insects carry large numbers of nematodes in their tracheae. They fly to the crown of healthy pines to feed on tender twigs. The beetles remove outer bark, and nematodes leave the insect to enter resin canals of the tree. Here they mature and reproduce, building large populations throughout the tree. While host tissues are moist, *B. xylophilus* reproduces rapidly. At favorable temperatures a complete generation is produced in 5 days. The tree dies from the infection during the summer and begins to dry; dying pines are invaded by fungi upon which the nematodes feed. Resistant fourth-stage juveniles (dauer-larvae) appear in the nematode population. Meanwhile, during summer, beetles oviposit on stressed trees in which nematode populations are building. The insect larvae tunnel into the wood and adults emerge the following spring. Dauerlarvae aggregate around developing pupae and enter the mesothoracic spiracles of newly molted adults. The nematodes continue their life cycle when beetles find a healthy tree. Figure 6.40 shows the nematodes in pine tissues and Fig. 6.41 is an SEM view of dauerlarvae in the beetle.

Many but not all species of pine are susceptible to nematode-induced pine wilt. In addition to transmission during feeding, transmission during oviposition also occurs. Thus felled logs in forests are sites of oviposition and serve as sources of beetles carrying *B. xylophilus*. Climate influences the occurrence of pine wilt. Trees in northern locations may be infected, but the nematode populations do not reach high levels, and pines usually

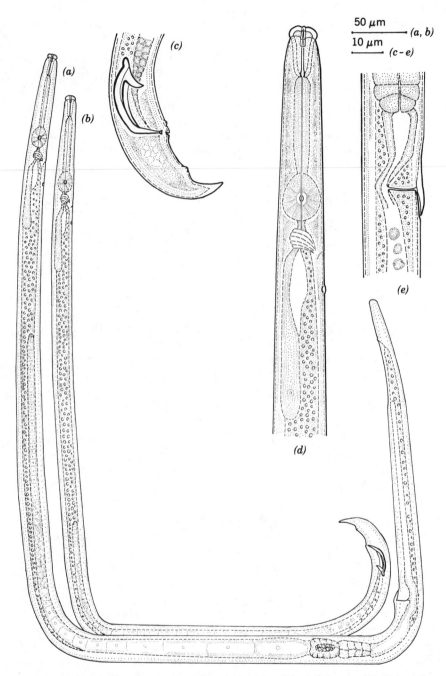

Fig. 6.39 Morphology of *Bursaphelenchus xylophilus*. (**a**) adult female; (**b**) adult male. (**a**) and (**b**) (×300); (**c**) male tail; (**d**) female, anterior portion; (**e**) female, vulva. (**c**) – (**e**) ×855). (From Y. Mamiya and T. Kiyohara. 1972. Description of *Bursaphelenchus lignicolus* n. sp. (Nematoda: Aphelenchoididae) from pine wood and histopathology of nematode infested trees. Nematologica 18:120 – 124, by permission of E. J. Brill.)

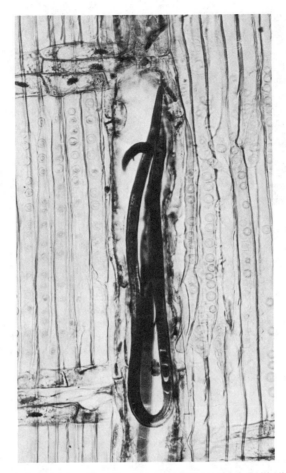

Fig. 6.40 *Bursaphelenchus xylophilus* in an axial resin canal. Epithelial cells are destroyed (radial section). (From Y. Mamiya. 1976. Pine wilting disease caused by the pinewood nematode, *Bursaphelenchus lignicolus* in Japan. JARQ 10:206 – 211, by permission of the author.)

survive. Nematode populations from different geographic locations in the United States may have different host ranges. In Japan, this nematode varies in tree-killing capacity.

Pathology

The first external symptom of infection observed within a few weeks after nematodes enter a tree is the reduction or absence of resin flow at wounds.

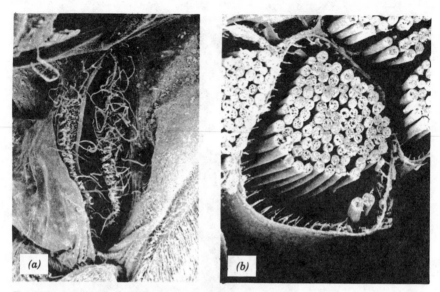

Fig. 6.41 Dauerlarvae of *Bursaphelenchus xylophilus* in metathoracic trachea of *Monochamus carolinensis.* (a) Scanning electron micrograph of spiracle showing nematodes at opening (×51); (b) SEM of trachea cut transversely to show Dauerlarvae (×380). (From E. Kondo, A. Foudin, M. Linit, M. Smith, R. Bolla, R. Winter, and V. H. Dropkin. 1982. Pine Wilt Disease, nematological, entomological, and biochemical investigations. Spec. Rep.— Mo., Agric. Exp. Stn. 282:1 – 56, by permission.)

Needles begin to lose their intense green and quickly become rust colored. Trees appear to die suddenly.

During feeding on twigs in the crown of a healthy tree, beetles strip the bark and expose the wood. Dauerlarvae enter resin canals at the wound to feed on cells lining the canals. The tree responds to internal wounds by secreting copious amounts of resin at each wound, and parenchyma cells near the area of damage die. Perhaps diffusible substances from the nematodes or from the damaged resin canals reach toxic levels. Resin canals forming a network of interconnected passages, as well as other avenues of transport, become plugged. *B. xylophilus* nematodes are highly active and reproduce quickly. Small juveniles move through vessels and soon invade the entire tree, from roots to top, feeding on nonwoody cells wherever they are. Transpiration declines and eventually ceases and the infected tree dies from lack of water. Some investigators believe that a toxin is produced that participates in the pathology. Destruction of cambium also contributes to death of affected trees. Once resin flow has halted the tree is doomed.

Interaction with Other Pathogens

There is no specific association of pine wilt nematodes with bacteria or fungi. Healthy trees defend themselves against bacteria and fungi, but when pine wilt caused by nematodes occurs, bacteria and fungi invade damaged tissues. Some investigators think that bacteria produce a toxin that kills trees, but experimental infection by nematodes in the absence of bacteria also kills. It is therefore highly likely that the nematode alone is the initial etiologic agent of pine wilt. Fungi are important because *B. xylophilus* maintains its population by feeding on fungi while the insect life cycle is progressing.

Epidemiology

Pine wilt results from the interaction of nematodes, fungi, beetles, and pines. Each organism responds to its environment and contributes to the final result. At the turn of the century, foresters in Japan reported that pines were dying from unknown causes in the vicinity of Nagasaki, a port in southern Japan. The malady spread slowly through southern Japan before World War II and rapidly thereafter. Following extensive unsuccessful efforts to determine the causal agent(s), *Bursaphelenchus* was found in affected trees, and pines died after inoculation with this organism.

In recent years pine forests of Japan have been severely damaged. The nematode is also present in North America, but does not cause widespread destruction. Why not? Japanese forests consisted of plantations of *Pinus densiflora* or *P. thunberghii*. Both are highly susceptible to pine wilt. The beetle vector, *Monochamus alternatus*, is efficient in transmitting the pathogen. Southern and central Japan are warm, moist, and subject to destructive storms. In the first half of the century, these man-made forests were important sources of wood used for fuel. Fallen logs were utilized and beetle populations remained low. Then the war drained labor from the forests, petroleum replaced wood as a source of energy, and labor moved to industry. Beetle populations zoomed as available oviposition sites increased. Pine wilt also spread rapidly through the forests. Major efforts to limit the disease failed and Japanese pine forests were totally destroyed in some areas. A recent review is published in Ref. (7).

Conditions in North American forests do not favor rapid spread of pine wilt. Insect vectors are not as efficient as the Japanese beetle, climate is not highly favorable for transmission, and tree stands usually contain mixed species. Pine wilt has killed some introduced pines in urban settings and occasional pines in forests of the United States. Now that large scale tree farms of pines exist in Australia, New Zealand, South Africa, and to some extent in the United States, this disease is a serious threat. Ecology of the involved organisms must be understood thoroughly to avoid major losses.

Pine wilt illustrates exquisite adaptations of a group of organisms interacting with each other to ensure survival. Such interactions occur among nematode parasites of animals and of humans. Each organism involved in pine wilt responds to seasonal changes so that beetles emerge when pines are undergoing a flush of growth. Nematodes must be ready to come aboard beetles at this time. As trees lose moisture after death, the nematodes feed on fungi that tap food reserves of dead pines. *B. xylophilus* reproduces on many fungi in the laboratory, thus indicating its ability to utilize different fungi in the pine. Dauerlarvae appear just when adult beetles are emerging, probably in response to large nematode populations and gradual drying of the wood.

This nematode appears to be native to North America. It is distributed over the whole continent from Texas to Manitoba and from California to the eastern seaboard. It affects many species of pine as well as a few related conifers, but does not usually reach overwhelming levels. *Pinus sylvestris* and *P. thunberghii*, introduced species, are more susceptible than native pines. *B. xylophilus* was probably introduced into Japan where it found highly favorable conditions.

Control

In Japan, pine wilt has been widespread in pine forests of *P. densiflora* and *P. thunberghii* (Japanese red and black pines). Removal of the insect vector population by insecticides would control the disease, but this is not feasible in forests. Highly valuable plantings of pines, such as shelter belts protecting vegetable production near the sea, have been kept free of pine wilt by regular spraying with insecticides from helicopters. Control through use of genetic resistance is also possible, and vigorous programs to produce resistant pines are under way. Some fungi pathogenic to insects are known; until now, it has not been practical to use these on a large scale. Because trees killed by the nematode usually contain populations of developing vectors, the beetles must be eliminated by spraying or burning. The introduction of certain nematicides into living trees also offers promise for control. Japanese forests are now being replanted to replace lost pines with other conifers.

Research

Pine wilt has been a major problem in Japan for several decades (6, 7). The disease is present in the United States, but is not a major problem.

In Japan *Monochamus alternatus* is the principal beetle vector of the nematode. In the United States *M. carolinensis*, *M.titillator*, and *M. scutellatus* are vectors.

Two groups have studied the role of toxins (1, 9). Bolla considers that toxin in trees acts as a phytoalexin, albeit not an effective one (2). Oku believes that bacteria are central in toxin production (10).

Populations of *B. xylophilus* vary. Proteins from a French population of pine wilt nematodes are different from those of a Japanese population (3). Avirulent strains inoculated into susceptible pines protect the trees from subsequent damage by virulent strains (4). As in other well-studied phytonematodes, populations of *B. xylophilus* from different geographic areas have different host ranges.

A major plant breeding effort shows promise for restoration of forests in Japan by use of resistant clones of pines (5).

Conclusion

Many species of phytonematodes feed on roots and damage trees, especially in nurseries. Nematodes also transmit pathogenic viruses to trees. But *B. xylophilus* and *Rhadinaphelenchus cocophilus* are the only species known to kill large trees. They probably evolved from mycophagous ancestors (Aphelenchoididae) associated with wood-boring insects. Their short life cycles and great mobility enable these pathogens to spread rapidly within their hosts.

Nematodes, insects, fungi, and pines form a specialized cycle within coniferous forests, part of the grand cycle that returns dead trees to soil. Energy from the sun is locked into cellulose and other polymers in trees. This energy flows into many organisms, including bacteria, fungi, nematodes, insects, and their predators. Pine wilt disease hastens the return of energy to the soil. The particular circumstances in Japan—the moist climate, yearly typhoons, and monocultures of susceptible pines— resulted in major economic losses. The trend toward cultivation of trees as a crop, seen in the extensive pine plantations of Australia, New Zealand, and elsewhere, poses severe threats of losses by *B. xylophilus*. Vigilance is essential to detect and eliminate the disease before it can spread.

References

1. Bolla, R., Shaheen, F., and Winter, R. E. K. 1984. Phytotoxin in *Bursaphelenchus xylophilus* – induced pine wilt. In: Proceedings of the United States-Japan Seminar: The Resistance Mechanisms of Pines Against Pine Wilt Disease, Publ. No. 116, (V. H. Dropkin, ed.). Ext. Publ. Univ. of Missouri, Columbia, pp. 119 – 127.

2. Bolla, R., Shaheen, F., and Winter, R. E. K. 1984. Effect of phytotoxin from nematode – induced pinewilt on *Bursaphelenchus xylophilus* and *Ceratocystis ips*. J. Nematol.
16:297 – 303.

3. de Guiran, G., Lee, M. J., Dalmasso, A. and Bongiovanni, M. 1985. Preliminary attempt
to differentiate pinewood nematodes (*Bursaphelenchus* spp.) by enzyme electrophoresis.
Rev. Nématol. 8:88 – 90.

4. Kiyohara, T. 1984. Induced Resistance in Pine Wilt Disease. In: Proceedings of the
United States-Japan Seminar: The Resistance Mechanisms of Pines Against Pine Wilt
Disease, Publ. No. 116, (V. H. Dropkin, ed.). Ext. Publ. Univ. of Missouri, Columbia,
pp. 178 – 186.

5. Kobayashi, F. 1984. Strategies for the Control of Pine Wilt Disease. In: Proceedings of
the United States-Japan Seminar: The Resistance Mechanisms of Pines Against Pine
Wilt Disease, Publ. No. 116, (V. H. Dropkin, ed.). Ext. Publ. Univ. of Missouri, Columbia, pp.171 – 177.

6. Mamiya, Y. 1984. Perspectives of pine wilt disease in Japan. pp.6- 13; and Behavior of
the pine wood nematode, *Bursaphelenchus xylophilus*, associated with the disease development of pine wilt. In: Proceedings of the United States-Japan Seminar: The Resistance Mechanisms of Pines Against Pine Wilt Disease, Publ. No. 116, (V. H. Dropkin,
ed.). Ext. Publ. Univ. of Missouri, Columbia, pp. 14 – 25.

7. Mamiya, Y. 1984. The Pine Wood Nematode. In: Plant and Insect Nematodes, (W. R.
Nickle, ed.). Dekker, New York, Chapter 16, pp. 589 – 626.

8. Mamiya, Y. and Kiyohara, T. 1972. Description of *Bursaphelenchus lignicolus* n. sp.
(Nematoda: Aphelenchoididae) from pine wood and histopathology of nematode infested
trees. Nematologica 18:120 – 124.

9. Oku, H. 1984. Biological activity of toxic metabolites isolated from pine trees naturally
infected by pine wood nematodes. In: Proceedings of the United States-Japan Seminar:
The Resistance Mechanisms of Pines Against Pine Wilt Disease, Publ. No. 116, (V. H.
Dropkin, ed.). Ext. Publ. Univ. of Missouri, Columbia, pp.110 – 118.

10. Oku, H., Shiraishi, T., Ouchi, S., Kurozumi, S., and Ohta, H. 1980. Pine wilt toxin, the
metabolite of a bacterium associated with a nematode. Naturwissenschaften 67:198 –
199.

RHADINAPHELENCHUS COCOPHILUS

Introduction

"Red ring" disease of coconut and oil palms, like pine wilt, is caused by
a nematode in association with an insect. Nematodes of the taxonomic
group to which *Bursaphelenchus* and *Rhadinaphelenchus* belong include
parasites of insects and inhabitants of insect burrows where they feed on
fungi.

Classification

Order Tylenchida; suborder Aphelenchina; superfamily Aphelenchoidea;
family Aphelenchoididae; subfamily Rhadinaphelenchinae.

Morphology

Both males and females are about 1 mm long, with a length/width ratio ranging from 60 to 180. Females have a prominent, sclerotized cephalic skeleton. The head is rounded, not offset, and slightly narrower than the body behind it. Stylets measure about 15 μm long and have small knobs. Long, narrow esophageal glands overlap the intestine dorsally. Excretory pore is just behind the nerve ring. The vulva is at two-thirds the body length and has a wide, thick dorsal lip. The gonad extends anteriorly and has an elongate postuterine sac reaching three-fourths the distance to the anus. There are four lateral incisures. The tail narrows to a rounded terminus and is 10 – 17 anal body widths long. Males resemble females in general body form. The head is about five-eighths as wide as the body behind it. A ventrally curved tail narrows gradually to a pointed terminus. Caudal alae reach to the tail tip (1). Morphology of the nematode and pathology of the disease are illustrated in Fig. 6.42.

Biology

Development of *Rhadinaphelenchus* in palms is arrested at a resistant juvenile stage. This enters larvae of a palm weevil, *Rhyncophorus palmarum*, through either the mouth or spiracles. Juveniles persist throughout the insect's development and are introduced during oviposition into soft tissues of palms. Nematodes may be carried also in bits of diseased tissue clinging to the insects. The life cycle is short, less than 10 days. Red ring disease occurs in islands of the Caribbean and in Central and South America. It has not yet been found elsewhere.

Pathology

R. cocophilus builds large populations in cortical tissues of roots and in intercellular spaces of stems and leaves. A characteristic red or red-orange ring of necrotic tissues (3 cm wide) appears in stems about 2.5 cm below the stem surface. Lower leaves become yellow, then brown; other leaves follow and the tree dies in a few months. The disease affects trees for two years before and after they come into bearing. A toxin has been reported from diseased trees. In some areas, the disease appears as "little leaf"— smaller than normal leaves in coconut and oilpalm.

Interaction with Other Pathogens

None reported.

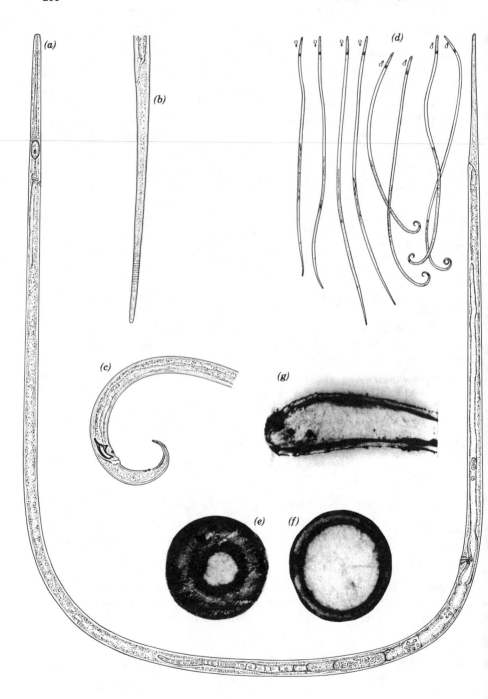

Control

Several methods of using insecticides against palm weevils can control red ring disease: sprays on trees, distribution in carriers placed in leaf axils, application to soil for systemic action. Sanitation is important; infested trees must be cut and burned to prevent spread of the infection. Treatment with arsenic to kill infested trees has also been used. Weevils have been trapped and killed by setting out wire baskets containing diseased coconut tissue treated with a fast-acting insecticide.

Research

African oil palms are reported to be resistant in Venezuela and elsewhere. However, African oil palms in a plantation in northern Brazil became vulnerable at about 5 yr of age. A local wild palm was the likely source of infestation. Early removal of infected trees, steam sterilization of these, and prevention of wounding controlled the disease (3).

An interesting genetic control of the nematodes in palm weevils operates in a proportion of the insects. A homozygous recessive allele stimulates production of an enzyme that lyses nematode juveniles in the hemocoele of weevil larvae. However, about 16% of weevil populations in some areas lack this defense mechanism. Weevils carrying nematodes are reduced in size and maintain the infection in palms (2).

References

1. Brathwaite, C. W. D. and Siddiqi, M. R. 1975. Commonwealth Institute of Helminthology, Descriptions of Plant-parasitic Nematodes, Set 5, No. 72.

2. Griffith, R. 1987. Red ring disease of coconut palm. Plant Disease 71:193 – 196.

3. Schuiling, M. and van Dinther, J. B. M. 1982. Red ring disease in the Paricatuba oilpalm estate, Para, Brazil. A case study. Z. Angew. Entomol. 91:154 – 169.

Fig. 6.42 *Rhadinaphelenchus cocophilus.* (a) Adult female (×400); (b) female tail (×560); (c) male tail (×560); (d) Adult males and females (×60); (e) – (g) Sections of coconut palm stems showing symptoms of red ring disease caused by *Rhadinaphelenchus cocophilus.* (e) and (f) Cross sections of stem showing red (dark) ring; (g) Longitudinal section of stem showing red (dark) lesions. (From C. W. D. Brathwaite, and M. R. Siddiqi. 1975. Commonwealth Institute of Helminthology, Descriptions of Plant- parasitic Nematodes, Set 5, No. 72, © Commonwealth Agricultural Bureaux, by permission.)

LONGIDORUS

Introduction

These are among the largest phytonematodes. In addition to damaging host roots directly, they transmit many troublesome viruses to plants. Their wide host range includes small fruits, deciduous trees, pasture crops, cereals, and corn. Hosts vary from one nematode species to the next.

Classification

Order Dorylaimida; suborder Dorylaimina; superfamily Dorylaimoidea; family Longidoridae; subfamily Longidorinae.

Morphology

Males are present in some species and unknown in others. Both sexes of *Longidorus* spp. are long, narrow nematodes. Length varies from about 4 to 10 mm or more, according to the species. The length/width ratio varies from about 70 to about 100. The head is not offset, and the body narrows at the lips to half to two-thirds the body width at the level of the guiding ring, a membranous fold in the wall of the esophagus. The stylet is in two parts: an anterior odontostyle (sclerotized spear) and a posterior odontophore (stylet extension, not sclerotized). Odontostyles range from 68 to 133 μm long and odontophores from 50 to 74 μm. The odontophore is not forked at its base and the odontostyle broadens at its base but does not have flanges. The guiding ring is far forward, in contrast to its position in *Xiphinema*. The esophagus is readily recognizable as a dorylaimoid type. It is usually looped at the junction between a narrow anterior portion and a wider, muscular, posterior bulb. The vulva is close to the midpoint of the body and gonads are paired, opposed, and reflexed. The cuticle appears smooth under the light microscope and has a series of pores in the lateral fields. The tail is dorsally convex and bluntly or roundly conoid (4). See Fig. 5.1(**b**) p. Figure 6.43 presents diagnostic features of the genus.

Biology

These ectoparasites resemble *Xiphinema* in their general biology. *Longidorus* spp. prefer sandy to medium loam soils. In some situations most of the population occurs at 60 to 90 cm deep in the soil, but in others the nematodes are closer to the surface. Individuals of a few species may live

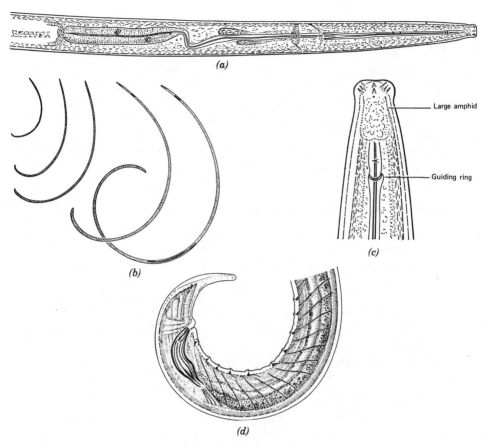

Large amphid

Guiding ring

(a)

(b)

(c)

(d)

Fig. 6.43 *Longidorus.* **(a)** *L. euonymus*, adult female, anterior portion (×250); **(b)** *L. euonymus*, stages in the life cycle (×16); From left to right: first-, second-, third-, and fourth-stage juveniles, and adult female; **(c)** *L. euonymus*, female, anterior end, lateral view (×790); **(d)** *L. nirulai*, adult male, tail (×325). **(a)** – **(c)** (From V. R. Mali and D. J. Hooper. 1973. Observations on *Longidorus euonymus* n. sp. and *Xiphinema vuittenezi* Luc et al., 1964 (Nematoda: Dorylaimida) associated with spindle trees infected with Euonymus mosaic virus in Czechoslovakia. Nematologica 19:459 – 467, by permission of E. J. Brill.) **(d)** (From M. R. Siddiqi. 1965. *Longidorus nirulai* n. sp., a parasite of potato plants in Shillong, India, with a key to species of *Longidorus* (Nematoda: Dorylaimoidea). Proc. Helminthol. Soc. Wash. 32:95 – 99, by permission of Helminthological Society of Washington, DC.)

for 3 – 5 years and some have survived for 2 years in soil stored in plastic bags.

Pathology

Longidorus nematodes feed at root tips and induce galls. Roots are stunted, tops grow poorly, yield is reduced. An individual nematode has been observed to remain in one position for as long as four days. Cells at the feeding site show cycles of increased synthesis and degeneration, finally ending in death. In certain combinations of *Longidorus* spp. and hosts, multinucleate hypertrophied cells appear close to the necrotic cells. Damage to roots is illustrated in Fig. 6.44.

Interaction with Other Pathogens

Necrotic gall tissue is invaded by bacteria, but no specific interaction with bacterial or fungal pathogens has been recorded. *Longidorus* transmits several viruses, including raspberry ring spot, brome grass mosaic, carnation ring spot, prunus necrotic ring spot, and tomato black ring. Virus – nematode relations are described in the discussion of *Xiphinema*. Transmission of virus by *Longidorus* is specific, in both nematode and virus. In experimental studies one species of nematode acquired two strains of virus, that appeared on the walls of the guiding ring and odontostyle. But only one strain was transmitted. Nematodes of this genus retain their capacity to transmit virus for only a few months; but because many weeds are hosts of both nematode and virus, transmission to crops continues as virus is readily reacquired.

Control

Weed control is important to break the cycle of virus transmission. Some cultivars of hosts resistant to virus are available but little effort to breed nematode-resistant crops has been made. Nematicides are used on high-value, perennial crops such as strawberries and raspberries.

Research

Research on nematode vectors of virus has increased in recent years. Taxonomy, histopathology, and details of virus transmission are subjects of interest. A species of *Longidorus* induces cycles of hypertrophy and hyperplasia in ryegrass, resulting in galls upon which the nematode feeds.

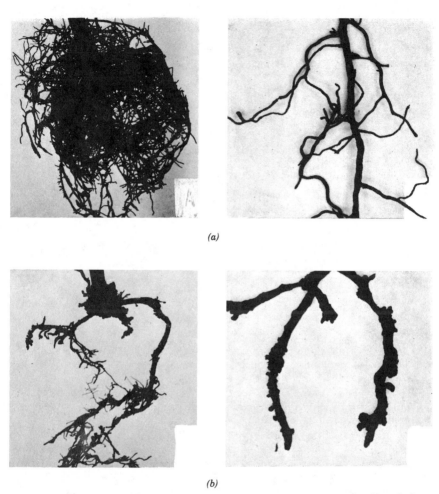

(a)

(b)

Fig. 6.44 Effect of *Longidorus africanus* on grape roots 9 months after inoculation. (a) Noninoculated control. (b) With *L. africanus*. Note: figures on left, low magnification; on right, close view of a portion of the root. (From E. Cohn. 1970. Observations on the feeding and symptomatology of *Xiphinema* and *Longidorus* on selected host roots. J. Nematol. 2:167 – 173, by permission.)

Galls collapse and bacteria invade the dead tissue. The entire cycle takes 10–12 days at 18°C (3). In celery roots attacked by *Longidorus* a cavity develops at the feeding site. Cells surrounding the cavity show a typical wound response as seen in ultrastructural changes (2). For details of feeding behavior, refer to Refs. (6) and (8).

Differences in preference of soil depth have been noted in a hops planting and in an adjacent pasture. *L. caespiticola* was concentrated at 40 – 50 cm depth under hops whereas *L.leptocephalus* was most dense at 10 – 20 cm. Barbez concluded that the pasture formed a focus of infestation for the hops (1).

Paralongidorus australis damages rice in Queensland, Australia. Rice is sown in soil with periodic application of water by irrigation. When plants are 10 – 15 cm high, fields are flooded and soon thereafter plants show symptoms of attack by nematodes: chlorosis and stunting, curling and necrosis of root tips. Severe loss of yield is the final result. Apparently these elongate nematodes, measuring up to 10.6 mm in length, do not move freely through dry soil. In flooded soil they move quickly to tips of growing roots (7). *P. australis* survives well under flooding, probably because rice roots contain adequate amounts of oxygen required by the nematodes.

A review of grapevine viruses transmitted by *Longidorus* spp. can be found in Ref. (5).

References

1. Barbez, D. 1982. The occurrence of virus vector nematodes in hops in Poperinge (Belgium) with notes on vertical and horizontal distribution, population structure and population density. Meded. Fac. Landbouwwet., Rijksuniv. Gent 47:741 – 755.
2. Bleve-Zacheo, T., Zacheo, G., Melillo, M. T., Lamberti, F. and Arrigoni, O. 1982. Ultrastructural response of celery root cells to *Longidorus apulus*. Nematol. Mediterr. 10:141 – 155.
3. Griffiths, B. S. and Robertson, W. M. 1984. Morphological and histochemical changes occurring during the life-span of root-tip galls on *Lolium perenne* induced by *Longidorus elongatus*. J. Nematol. 16:223 – 229.
4. Hooper, D. J. 1973, Commonwealth Institute of Helminthology, Descriptions of Plant-parasitifc Nematodes, Set 2, No.30: *Longidorus elongatus*; see also Set 5, No. 63: *Longidorus caespiticola*; Set 5, No. 67: *L. macrosoma*; Set 6, No. 88: *L. leptocephalus*.
5. Lamberti, F. 1978. Nematode vectors of grapevine viruses and their spread and control. Monogra. Instit. Nac. Investig. Agraria CNR. 18:239 – 246.
6. Robertson, W. M., Trudgill, D. L. and Griffiths, B. S. 1984. Feeding of *Longidorus elongatus* and *L.leptocephalus* on root-tip galls of perennial ryegrass (*Lolium perenne*). Nematologica 30:222 – 229.
7. Stirling, G. R. and McCullough, J. S. 1984. *Paralongidorus australis* n. sp. (Nematoda: Longidoridae), causing poor growth of rice in Australia. Nematologica 30:387 – 394.
8. Wyss, U. 1981. Ectoparasitic root nematodes: feeding behavior and plant cell responses. In: Plant Parasitic Nematodes, Vol. 3, (B. M. Zuckerman, ed.). Academic Press, New York, pp. 325 – 351.

XIPHINEMA

Introduction

These large phytonematodes damage plants directly and also act as vectors of viruses. Although a species may have a wide host range, individual species have distinctive sets of hosts. Many woody plants, such as fruit and other trees, are good hosts. Grasses and other pasture plants are also hosts. *Xiphinema* spp. are suitable for experimental manipulation and will probably be good models for investigating how nematodes survive and what they inject into plants.

Classification

Order Dorylaimida; suborder Dorylaimina; superfamily Dorylaimoidea; family Longidoridae; subfamily Xiphineminae.

Morphology

Xiphinema spp. are long, slender ectoparasites. Males are present in some, but not in all species. Length of adults ranges from 1.7 to 5 mm or more and the length/ width ratio from 50 to 75. The lip region is hemispherical, not set off from the body. The stylet is in two parts: an odontostyle (sclerotized spear) and an odontophore (stylet extension, not sclerotized). A guiding ring encircles the stylet close to its base. The stylet ranges from about 70 to 150 μm long and is forked at the base; the stylet extension has three large basal flanges and ranges from 45 to 85 μm long. The esophagus is a long, narrow cylinder, abruptly expanded posteriorly into a cylindrical bulb. This is half as wide as the body width at its base and two and a half times as long. Esophageal glands are within the bulb. When the stylet is retracted, the anterior esophagus is looped where it joins the expanded basal bulb to allow for stylet thrusting. The vulva is at the midpoint of the body or anterior to it, and there are usually two gonads, one extending anteriorly and another posteriorly. In some species, the anterior gonad may be incompletely developed or absent. Males, where present, have two testes. The cuticle appears smooth under the light microscope and has a series of lateral pores along the body. The tail is conoid or broadly rounded with a blunt ventral or terminal peg (8). Two illustrations (Figs. 6.45 and 6.46) show the important diagnostic characters of this genus.

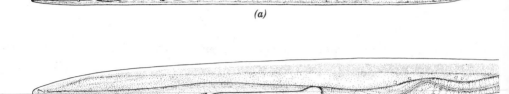

(a)

(b)

Fig. 6.45 *Xiphinema orthotenum* (**a**) Entire adult female (×66); (**b**) anterior of adult female (×240) (From E. Cohn and S. A. Sher, 1972. A contribution to the taxonomy of the Genus *Xiphinema* Cobb, 1913. J. Nematol. 4:36 – 65, by permission.)

Biology

Individuals of some species have great longevity; their life cycle may be 3 years from egg to egg. Nematodes of some species can survive for up to 3 years in soil without host plants. These large nematodes are sensitive to low rates of oxygen diffusion. They prefer light- or medium-textured soils.

Pathology

Dramatic reductions in root growth result from attack by *Xiphinema*, as shown in Fig. 6.47. The long flexible stylet penetrates and kills cells of the cortex. An individual may remain immobile for more than a day, feeding intermittently. Cells adjacent to the area of necrosis enlarge, and a gall results, especially at root tips. Nuclei of affected cells enlarge, and multinucleate cells appear. Extensive root damage results in reduced top growth.

Interaction with Other Pathogens

Bacteria and probably fungi invade root lesions induced by *Xiphinema*. However, the most important interaction is that these nematodes transmit pathogenic viruses to plants. They acquire virus particles during feeding on viruliferous plants. Acquisition time may be as short as a few minutes. Electron micrographs show particles on the lumen wall from the odontophore to the end of the basal bulb. When a viruliferous nematode subsequently feeds on a nonviruliferous plant, some virus particles are released

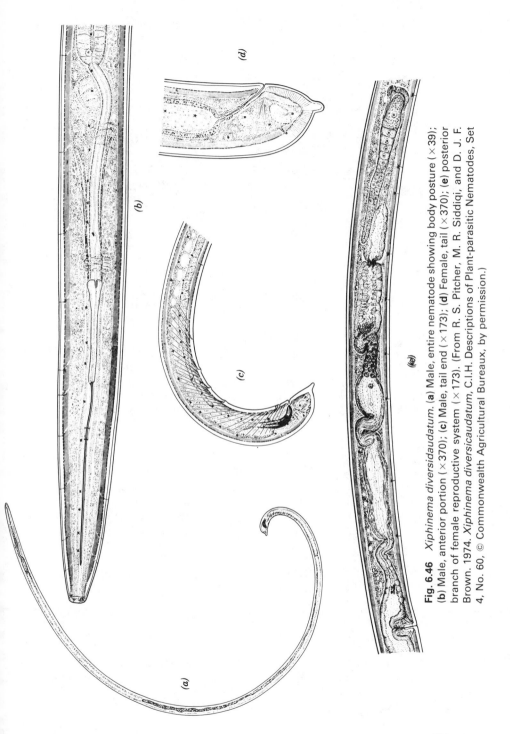

Fig. 6.46 *Xiphinema diversidaudatum.* (**a**) Male, entire nematode showing body posture (×39); (**b**) Male, anterior portion (×370); (**c**) Male, tail end (×173); (**d**) Female, tail (×370); (**e**) posterior branch of female reproductive system (×173). (From R. S. Pitcher, M. R. Siddiqi, and D. J. F. Brown. 1974. *Xiphinema diversicaudatum,* C.I.H. Descriptions of Plant-parasitic Nematodes, Set 4, No. 60, © Commonwealth Agricultural Bureaux, by permission.)

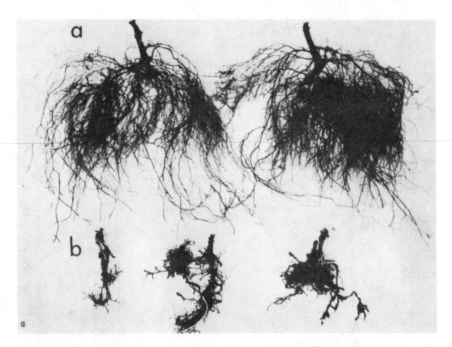

Fig. 6.47 Effect of *Xiphinema bakeri* on raspberry roots. (a) Roots of plants grown in steamed soil for 12 weeks; (b) Roots of plants grown in soil with 500 nematodes/pot. (From F. D. McElroy. 1972. Studies on the host range of *Xiphinema bakeri* and its pathogenicity to raspberry. J. Nematology 4:16 – 22, by permission.)

and injected into the host. One nematode may transmit virus to more than one host plant. A viruliferous *Xiphinema* may continue to infect plants for weeks when hosts are available and also after months of starvation. Virus particles are lost during molting as the esophageal lining is discarded and the nematode loses its ability to transmit. These nematodes show marked specificity of virus transmission. A population of *Xiphinema* may transmit a particular strain of virus but another population of the same species may transmit other strains.

Control

Crop rotation is feasible because host ranges of *Xiphinema* spp. are well defined. However, they must be continued for more than a few years because these nematodes may survive for years in the absence of a host. But ability to transmit virus is eventually lost. Therefore it is possible to free a field from viruliferous nematodes by avoiding use of virus- susceptible crops. Within a year or two, the remaining nematode population will be free of virus and virus-susceptible plants may be grown.

Nematicides depress populations of *Xiphinema* free in soil. However, endoparasites within roots are often protected from lethal concentrations; chemical control of nematode pests in vineyards requires deep placement of nematicides to reach grape roots that remain in soil after vines are rogued.

Research

Because plant diseases caused by viruses are important, much research has been done on nematodes that transmit virus. These include *Xiphinema, Longidorus, Trichodorus* and closely related genera. For a general review consult Ref. (5). The original demonstration that nematodes are vectors of virus is a model of scientific investigation (6). Large pots of plants were maintained on benches aboveground in vineyards showing fan leaf disease, known to be caused by a virus. Healthy rooted grape cuttings remained healthy in soil free of *Xiphinema index*. A cutting grown in a pot of soil containing a diseased cutting (grape fan leaf) became infected only when *X. index* was present. This work stimulated others to explore the role of nematodes in transmission of virus to plants. Brown (2) demonstrated specificity of transmission by crossing a population of *Xiphinema* from Scotland with one from Italy. The Scottish nematodes readily transmitted a British strain of strawberry latent ring spot virus and another of arabis mosaic virus whereas the Italian nematodes transmitted these viruses only with difficulty. Progeny of reciprocal crosses showed the influence of both parents. The author suggested that some heritable characteristic of the cuticular lining of the esophagus controls specific transmission of virus strains.

The ecology of *Xiphinema* is of practical importance for control of virus transmission. Populations tend to remain localized close to host plants although they are capable of moving through soil. *X. diversicaudatum* migrated through a sandy loam soil at the rate of 7 cm/month vertically and 10 cm/month horizontally (10). Females survived for 60 months on strawberry and produced about 200 eggs during this period (3).

Cellular response to feeding by *Xiphinema* continues to attract investigators (1, 4, 8). Sophisticated microanalytic techniques have been applied to newly formed galls of *Ficus* and to the nematodes that induced them (7). Proteins in galls are the same as those in unaffected roots. The authors detected glycoproteins in esophageal bulbs of nematodes, which they believe are part of the secretions introduced into plants during feeding.

The surface of *Xiphinema* has the same types of carbohydrate residues as found in some other genera, but the locations of particular residues are distinctive. Sialic acid residues are confined to certain spots on the head whereas galactose N-acetyl-D-galactosamine residues are present along the entire body wall (9).

Conclusion

Transmission of pathogenic viruses by nematodes to plants is another example of a complex adjustment among three kinds of organisms. Although many phyto-nematodes feed by extracting cell contents, thereby taking up virus particles from viruliferous plants, only three groups of species, all belonging to the Dorylaimida, are vectors of virus. The esophageal lining must have particular characteristics to permit virus particles to maintain their position in the anterior portion of the alimentary canal. And a mechanism for the release of some particles must operate during feeding on uninfected plants. Feeding must not destroy cells quickly or virus replication will be prevented. In addition, plants must be hosts for the nematodes and for the virus. Evolution of such associations is probably the result of a long history including many failures.

References

1. Bleve Zacheo, T. and Zacheo, G. 1983. Early stage of disease in fig roots induced by *Xiphinema index*. Nematolog. Mediterr. 11:175 – 187.

2. Brown, D. J. F. 1986. Transmission of virus by the progeny of crosses between *Xiphinema diversicaudatum* (Nematoda: Dorylaimoidea) from Italy and Scotland. Rev. Nématol. 9:71 – 74.

3. Brown, D. J. F. and Coiro, M. I. 1983. The total reproductive capacity and longevity of individual female *Xiphinema diversicaudatum* (Nematoda: Dorylaimida). Nematol. Mediterr. 11:87 – 92.

4. Griffiths, B. S. and Robertson, W. M. 1984. Nuclear changes induced by the nematode *Xiphinema diversicaudatum* in root-tips of strawberry. Histochem. J. 16:265 – 273.

5. Harris, K. F. 1981. Arthropod and nematode vectors of plant viruses. Annu. Rev. Phytopathol. 19:391 – 426.

6. Hewitt, W. B., Raski, D. J., and Goheen, A. C. 1958. Nematode vector of soil-borne fanleaf virus of grapevines. Phytopathology 48:586 – 595.

7. Poehling, H. M. & Wyss, U. 1980. Microanalysis of proteins in the aseptic host-parasite system: *Ficus carica-Xiphinema index*. Nematologica 26:230 – 242.

8. Siddiqi, M. R. 1973. Commonwealth Institute of Helminthology, Descriptions of Plant-parasitic Nematodes, Set 2, No. 29: *Xiphinema americanum*; see also Set 3 No. 45: *X. index*; Set 4 No. 60: *X. diversicaudatum*.

9. Spiegel, Y., Robertson, W. M., Himmelhoch, S., and Zuckerman, B. M. 1983. Electron microscope characterization of carbohydrate residues on the body wall of *Xiphinema index*. J. Nematol. 15:528 – 534.

10. Thomas, P. R. 1981. Migration of *Longidorus elongatus, Xiphinema diversicaudatum* and *Ditylenchus dipsaci* in soil. Nematolog. Mediterr. 9:75 – 81.

TRICHODORUS AND PARATRICHODORUS

Introduction

The importance of ectoparasites first came to attention during trials of fumigants in which nematicides were accidentally omitted in some rows of a heavily infested field. Obvious damage to plants in these rows led to discovery of high numbers of *Trichodorus* spp. associated with poor growth. Roots affected by these organisms have a distinctive appearance called "stubby root." *Trichodorus* spp. parasitize a wide range of hosts, including many of economic importance: sugar beet, peas, potato, strawberry, cabbage, fruit trees, corn, cotton, wheat, and many others.

Classification

Order Dorylaimida; suborder Diphtherophorina; family Trichodoridae.

Morphology

Males are usually abundant in the genus *Trichodorus*. In the related genus *Paratrichodorus*, some species have males; in others they are rare. Nematodes of the Trichodoridae are plump and cigar-shaped, ranging from 0.5 to 1.5 mm long. The length/width ratio is between 15 and 33. The rounded head is not offset. The stylet is a ventrally curved, elongate tooth, attached to the dorsal pharyngeal wall and its length varies from 40 to 50 μm. There are two parts: an anterior outer spear and a posterior inner spear. The esophagus is a narrow cylinder, broadening posteriorly into a pear- or spoon-shaped bulb that abuts the intestine in *Trichodorus* spp. or overlaps it ventrally in *Paratrichodorus* spp. The vulva is just posterior to the equatorial position; gonads are paired, opposed, with reflexed ovaries. The testis is single. *Trichodorus* males have no bursa, and a bursa is present in *Paratrichodorus*. The cuticle of *Trichodorus* does not swell upon fixation, but that of *Paratrichodorus* does. The tail is broadly rounded, and the anus is close to the terminus (3). Morphology of these nematodes is shown in Fig. 6.48.

Biology

Trichodorus and *Paratrichodorus* are found mostly in light to medium textured sandy soils, often at fairly deep levels, down to 60 cm. They are

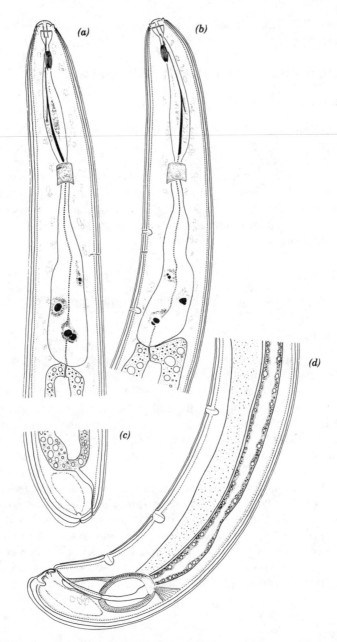

Fig. 6.48 *Trichodorus coomansi* (**a**) anterior region, female; (**b**) anterior region, male; (**c**) tail region, female; (**d**) posterior region, male. (**a**), (**b**), (**d**) ×570; (**c**) ×745. (From D. DeWaele and E. Carbonell. 1982. Two new species of *Trichodorus* (Nematoda: Diphtherophorina) from Africa. Nematologica 28:387 – 397, by permission of E.J. Brill.)

more sensitive to desiccation than most phytonematodes, but when soils dry gradually, they can survive low moisture. These ectoparasites aggregate at or just behind root tips. To feed, the nematode turns its head at right angles to the surface of an epidermal cell, and the stylet thrusts very rapidly (10 thrusts/sec have been recorded). Individual nematodes remain in one position for a few minutes. Cytoplasm of a penetrated cell aggregates and the nematode imbibes cell contents by peristaltic movements of the esophagus. Some species secrete a feeding tube that coagulates around the stylet and is left in position when the nematode withdraws (5).

Pathology

Roots attacked by *Trichodorus* or *Paratrichodorus* turn brown and cease to elongate. Numerous side roots develop, each of which is attacked again, giving rise to the symptom called "stubby root." Plants with severe stubby root grow poorly. Large populations of these nematodes have caused complete crop failure of corn in some fields of southern United States. Figure 6.49 shows "stubby root" of blueberry.

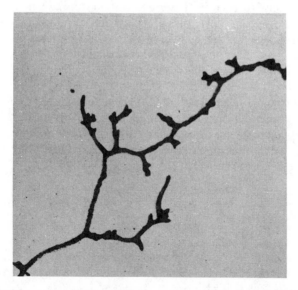

Fig. 6.49 Stubby root symptom of blueberry attacked by *Paratrichodorus christiei.* (From B. M. Zuckerman. 1962. Parasitism and pathogenesis of the cultivated highbush blueberry by the stubby root nematode. Phytopathology 52:1017 – 1020, by permission of American Phytopathological Society.)

Interaction with Other Pathogens

Trichodorus and *Paratrichodorus* spp. are vectors of tobacco rattle, and pea early browning viruses. Virus particles adhere to the lining of the alimentary canal from the stoma to the intestine. Virus may be retained in the nematodes for as long as 3 years in fallow soil lacking hosts.

The influence of *Trichodorus* spp. on concurrent infection by fungi and bacteria is not well known. One author found that *Paratrichodorus christiei* increased the incidence of *Verticillium albo-atrum* on tomato. Presence of the fungus did not affect reproduction of the nematode (1).

Control

Trichodorid nematodes are highly susceptible to nematicide treatment of soils and chemical control is widely used where economically advantageous. However, these nematodes are often the first to reinvade fumigated fields, reaching much higher population levels than existed before nematicide treatment.

Research

For a comprehensive review of the association between nematodes and plant viruses consult Ref. (4). Viruses transmitted by *Xiphinema* spp. and *Longidorus* spp. are known as NEPO viruses in reference to their poly-hedral shape. They include several different kinds, named for the diseases they cause: arabis mosaic, cherry leaf roll, grapevine fanleaf, raspberry ring spot, tomato black ring, and some others. NEPO viruses are also transmitted through seed and pollen (4). Those transmitted by *Trichodorus* spp. and *Paratrichodorus* spp. are called TOBRA viruses in reference to their tubular shape. These include pea early browning and tobacco rattle. In most cases nematodes are not the sole agents of virus transmission. Keys to species are available in Ref. (2).

References

1. Conroy, J. J. and Green, R. J., Jr. 1974. Interactions of the root knot nematode *Meloi-dogyne incognita* and the stubby root nematode *Trichodorus christiei* with *Verticillium albo-atrum* on tomato at controlled inoculum densities. Phytopathology 64:1118 – 1121.
2. Decraemer, W. 1980. Systematics of the Trichodoridae (Nematoda) with keys to their species. Rev. Nématol. 3:81 – 99.
3. Hooper, D. J. and Siddiqi, M. R. 1972. Commonwealth Institute of Helminthology, Descriptions of Plant-parasitic Nematodes, Set 1, No. 15: *Trichodorus primitivus*; see also Set 4, No. 59: *T. similis*; set 5, No. 69: *Paratrichodorus christiei*; set 6, No. 86: *T. viruliferus*.

4. Taylor, C. E. 1978. Plant-parasitic Dorylaimida: Biology and virus transmission. In: Plant Nematology, (J. F. Southey, ed.). Ministry of Agriculture, Fisheries and Food, GD1. H. M. Stationery Office, London, Chapter 11, pp. 232–843.

5. Wyss, U. 1971. Der Mechanismus der Nahrungsaufnahme bei *Trichodorus similis*. Nematologica 17:508 – 518.

Seven

PATHOLOGY

Introduction
Parasitic Habit
Pathology of Cells
Pathology of the Whole Plant
Mechanisms of Pathogenesis
References

INTRODUCTION

The most striking effect of nematode infection is a general reduction in growth. A patch of poorly growing plants in an otherwise healthy crop is often the first sign of nematode problems. Affected plants usually wilt earlier than healthy ones. In row crops, such as soybeans or potatoes, the healthy crop forms a canopy over the soil, excluding weeds. But nematode-infected plants fail to form a canopy and weeds often grow profusely in affected areas.

Associations between nematodes and plants illustrated in Chapter 6 display a great variety. At one extreme, they are almost casual; at the other, nematodes induce major responses in cells upon which they feed. These associations will be considered in several ways:

1. According to parasitic habit of nematodes, whether they remain external to plant tissues or enter.
2. At the cellular level.
3. At the level of the whole plant.
4. According to the mechanism of pathology.

PARASITIC HABIT

Table 7.1 is a summary of parasitic habits of the major phytonematodes.

PATHOLOGY OF CELLS

Cellular pathologies induced by nematodes that feed on outermost tissues range from no visible effect to necrosis. *Paratylenchus* has been observed to remain in position, attached to a root by its stylet, feeding on one cell for several days without killing it. *Helicotylenchus* spp. induce local lesions at feeding sites in the cortex without affecting neighboring cells. Damage to roots is slight. *Belonolaimus longicaudatus*, apparently a voracious

Table 7.1   Parasitic Habit of Phytonematodes[a]

Ectoparasites

Nematodes that remain outside the plant and penetrate with only a small portion of their body:
a. Feeders on surface tissues: *Paratylenchus, Trichodorus, Tylenchorhynchus*
b. Feeders on subsurface tissues: *Belonolaimus, Criconemoides, Helicotylenchus, Hemicycliophora* (G), *Hoplolaimus, Longidorus* (G)[b], *Rotylenchus, Scutellonema, Xiphinema* (G)

Endoparasites

Nematodes which enter plant tissues completely or with a large portion of their body:
c. Migratory
   In roots of herbaceous plants: *Hirschmaniella, Pratylenchus, Radopholus*
   In stems and leaves: *Ditylenchus* (G)
   In buds and leaves: *Anguina*(G), *Aphelenchoides*
   In trees: *Bursaphelenchus, Rhadinaphelenchus*
d. Sessile, partly within roots: *Heterodera, Rotylenchulus, Tylenchulus*
e. Sessile, entirely within roots: *Nacobbus* (G), *Meloidogyne* (G)

[a]This classification is not absolute, because the particular parasitic relationship between a nematode and its host depends on both partners. For example, on some hosts, *Meloidogyne* nematodes may be partly embedded in a root, but on most hosts, they are entirely within plant tissues. *Heterodera* spp. develop entirely within roots; swollen bodies of females break through root tissues to the surface. Some species of *Hoplolaimus* enter entirely into root cortex.
[b](G) indicates that galls are usually present.

feeder, destroys large numbers of cells. Relatively few nematodes cause severe root damage. When the parasite destroys tissues destined to develop into leaves and flowers, as in the case of *Aphelenchoides*, the mature plant has malformed structures.

Migratory endoparasites, such as *Pratylenchus* and *Radopholus* cause extensive necrotic lesions as they move through cortex cells. Eggs deposited there hatch into juveniles and the lesion grows as the population increases. Eventually the nematodes move to fresh tissues or leave the root when bacteria or fungi invade.

Nematodes also affect cells beyond those directly destroyed by feeding. For example, citrus roots respond to *Radopholus similis* by forming galls consisting of hyperplastic pericycle tissue. *Longidorus* spp. insert their long stylets into root tips of host plants. Galls form just behind tips but vascular differentiation extends almost to the end of the root and elongation stops. *Trichodorus* spp. feed on root epidermis, including cells at the base of root hairs, without penetrating more deeply into cortex. Vascular differentiation continues almost to the root tip; elongation halts. Side roots emerge—the nematodes feed on these. The resulting stubby roots and severely inhibited plant growth are characteristic of infections with *Trichodorus* and *Paratrichodorus*.

Modification of cell structure, and presumably of cell function, without loss of integrity are characteristic of many plant – nematode associations. Not only sedentary endoparasites such as *Heterodera* and *Meloidogyne*, but also an ectoparasite, *Xiphinema*, induce host cells to respond to feeding in this way. Observations of cell alterations during feeding are difficult to make because plant roots are too thick and not translucent enough for good images at high magnification. By using freshly germinated fig seedlings and *Xiphinema index* under aseptic conditions, Wyss produced good plastic-embedded sections that permit detailed analysis of accurately timed events (15). The nematodes penetrate root tips to feed on cells in the zone of elongation. A few cells are killed and root growth halts within 12 hr after feeding commences. These cells are surrounded by multinucleate, hypertrophied cells with up to 20 nuclei. Distal to these are uninucleate hypertrophied parenchyma cells. Walls of multinucleate cells are thin and do not have the characteristic wall ingrowths of transfer cells. Their cytoplasm is dense, with many small vacuoles and large ameboid nuclei. We may assume that multinucleate cells synthesize products upon which the nematodes feed. In older galls, multinucleate cells are empty, presumably drained by *Xiphinema*.

The highly specialized multinucleate giant cells induced by *Meloidogyne* and syncytia induced by *Heterodera*, as well as giant uninucleate cells of certain other infections, indicate that each kind of sedentary endoparasite introduces a set of stimuli to which the plant responds in its characteristic way.

Recently Wyss and associates have used differential interference contrast microscopy together with video contrast enhancement to produce cine films of nematodes feeding within small roots under aseptic conditions (16). These permit frame-by-frame analysis of nematode behavior and plant response.

PATHOLOGY OF THE WHOLE PLANT

Roots have many functions. They anchor plants, take up water and minerals for export to upper parts, and convert absorbed nitrates to reduced nitrogen for utilization in the synthesis of protein and other compounds. Root tips produce certain plant growth regulators that are distributed to the whole plant. Products of photosynthesis move down from leaves and are exported back as compounds used in growth of upper plant parts. By this process, roots help to regulate rates of photosynthesis. In legumes and certain other plants, roots are the site of nitrogen fixation by symbiotic bacteria. Young plants forming roots and leaves are especially vulnerable to nematode attack because damage at this stage prevents normal plant development. Older plants have already accumulated reserves of nutrients and have more fully formed structures. They can often tolerate moderate nematode attack without serious damage.

1. Nematode infection of roots may alter mineral uptake. Is the alteration specific to each kind of nematode and plant or are the changes in mineral uptake general responses to root damage? Two different populations of *Heterodera glycines* have different effects on mineral uptake by soybeans. In replicated experiments, one population stimulated zinc uptake; the other did not (D. Blevins, 1987, personal communication). Calcium uptake may be enhanced by nematode infection. Destruction of the endodermis in oats by *H. avenae* may remove this barrier to Ca^{2+} uptake (11). Increased Ca^{2+} content has also been found in potato plants infected with *Globodera rostochiensis*, and in bitter almond roots infected with *Meloidogyne*. However, changes in other minerals are inconsistent. In comparison with healthy plants, roots of bitter almonds with *Meloidogyne* have higher concentrations of N, K, Mn, and Cu and lower Fe (9). Lower K and PO and higher Na and Mg were recorded in leaves of beets with *H. schachtii* (1). By microanalysis of vacuole contents, these investigators found higher potassium concentrations close to the nematodes in roots of infected beets, but lower in the vacuoles of leaf cells. Thus the partitioning of potassium is altered, although total uptake is increased in infected roots.

2. Nematode infection also causes alterations in organic compounds. Carbohydrate content of roots with *Meloidogyne* is reduced. This has been found in bitter almond, tomato, and pepper, and also in alfalfa with *Ditylenchus dipsaci*. Nucleic acid content is increased and composition is altered in infections of various plants with *Meloidogyne*. Total nitrogen increases in infections with *D. dipsaci*, and in galls of *Meloidogyne*. Amino acid ratios change in infections of oil radish and rape with *H. schachtii*, in onions with *D.dipsaci*, in bitter almond with *Meloidogyne*, and in fig with *Xiphinema*.

Roots of tomato infected with *Meloidogyne incognita* exude increased amounts of carbohydrates as well as other compounds (14). This, together with the report cited above of changes in the distribution of potassium in beets infected with *H. schachtii*, indicates that certain nematodes cause changes in permeability of membranes and consequent increased leakage from roots. As shown in the interaction between *Meloidogyne* and *Rhizoctonia solani*, this increased exudation results in enhanced fungal pathogenesis. Important references for this subject are (8), (10), and (14). Accumulation of starch in nodules of roots with *Nacobbus* provides further evidence that nematodes disturb carbohydrate metabolism.

3. *Meloidogyne* and *Heterodera* reduce nitrogen fixation by symbiotic bacteria. Three genetic systems interact here: nematodes, plants, and bacteria. *H. glycines* of a particular kind (race 1) suppresses binding of rhizobia to soybean roots by altering production of soybean lectin, a glycoprotein that affects bacterial attachment sites (5). That these effects are general throughout a root is shown by split-root experiments with nematodes in one part of a root system. Nodulation is suppressed on both root halves in heavy infections, and when the portion with nematodes is removed, the remaining portion nodulates normally (6, 7). *Scutellonema* and *Aphasmatylenchus*, as well as *Pratylenchus*, reduce nitrogen fixation on peanut. We may expect that other nematodes inhibit symbiotic nitrogen fixation. In addition, *Hirschmaniella* sp. reduces nonsymbiotic nitrogen fixation in the rhizosphere of infected rice plants (12).

4. Nematodes affect enzyme activities in plants. Sucrase and invertase increase in galled roots of eggplant with *M. javanica*. The nematodes themselves have a sucrose hydrolase (3). Tomatoes with *M. incognita* have increased superoxide dismutase and reduced photosynthesis, related no doubt to the disturbance in root function (17).

5. Demand for nutrients varies from time to time and from place to

place during growth of a plant. Activated buds of all kinds are regions of rapid growth. Fruit formation requires massive movement of metabolites and water for rapid synthesis of new cells and the storage of cell contents. Movement of solutes in plants from regions of production (leaves) to places of utilization undergoes great changes as a plant progresses from seedling to growing plant to flowering and seed or fruit production. Nematode activities alter these patterns by creating new metabolic sinks and by disturbing normal growth of tissues. We see the effects of this alteration in lowered crop yields.

These results suggest that nematode infection induces profound alterations of metabolism in plants. Galls are the sites of high metabolism, especially in pathways leading to the synthesis of protein. Figure 7.1 illustrates the accumulation of more radioactive C in galls than in unaffected stem tissue.

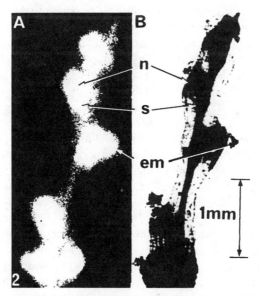

Fig. 7.1 Root of *Lycopersicon esculentum* parasitized with 7-week-old *Meloidogyne javanica* and harvested 5 days after exposure to $^{14}CO_2$. (a) Autoradiograph of freeze-dried root showing accumulation of labeled photosynthate in region of nematode (n), egg mass (em), and syncytia (s). (b) Root removed from surface of X-ray film and stained to show positions of nematode, egg mass, and syncytia. (×20). (From A. F. Bird, and B. R. Loveys. 1975. The incorporation of photosynthates by *Meloidogyne javanica*. J. Nematol. 7:111 – 113, by permission.)

MECHANISMS OF PATHOGENESIS

A few nematodes probably introduce growth regulators with their secretions or alter the host's own growth regulators. *Meloidogyne* juveniles probably introduce growth stimulants into root tissues as they enter. A gall is induced within 1 day after penetration, and its size is proportional to the number of nematodes in the gall. Enhanced ethylene production increases gall size (4). Gall size is also related to host and to nematode species. In contrast to the localized growth stimulation induced by *Meloidogyne*, *Ditylenchus dipsaci* has a more general effect on host plants. Growth is stunted, and stems swell and may be deformed. These symptoms may result from introduction of growth regulators by the nematode, from release of products of cell disintegration, or from disturbance of a plant's mechanism of growth hormone regulation. Inhibition of elongation may be a sign of inhibition of the host's own growth regulator balance (13).

Phytonematodes have relatively enormous esophageal glands in comparison with species that feed on bacteria. We assume that secretions from these glands are involved in induction of pathological changes in hosts. But critical proof of this is lacking. Three gland cells empty into the anterior alimentary canal of *Meloidogyne*. Ducts from two subventrals lead to the lumen of the median bulb; the duct of the dorsal gland extends anteriorly to the esophageal lumen just behind the stylet base. Contents of subventral gland ducts of infective juveniles change appearance during penetration. Perhaps secretions from these glands are important during early invasion of roots. Secretions from dorsal glands are probably injected into plant cells. Adult female *Meloidogyne* dissected from roots exude from their stylets a viscous material containing peroxidase and basic proteins. All sessile parasites that induce the most dramatic cell alterations have a large dorsal gland leading to the anterior region of the esophagus. Electron micrographs of cytoplasm close to the stylet orifice of *Rotylenchulus* furnish evidence that phytonematodes introduce gland secretions into cells. Modern analytical techniques now being applied to nematode stylet exudates should tell us how nematodes persuade plants to serve them.

Cell separation is a prominent aspect of infections with *Ditylenchus dipsaci*. Host tissues become fragile, dry rapidly, and develop large regions of loosely connected cells that disintegrate into cavities. Good evidence for the release of cell-wall-degrading enzymes from the nematodes has been obtained. Induction of adaptive cell changes, such as those in *Heterodera* and *Meloidogyne* infections, must be the result of secretions from the nematode that regulate cell function. In addition, part of the alteration of these feeding sites reflects the plant's system of responding to any metabolic sink, that is, a place of high rates of utilization of substrates for metabolism and growth. Plants normally develop specialized cells, called **transfer cells** for short-distance transport of solutes. They occur along borders of vessels in leaf traces, nectaries, salt glands, between xylem and phloem in leaf parenchyma, in nurse cells of ovules, and in other locations where great quantities of solutes are moving. Their morphology is specialized: the central vacuole is lost, the nucleus is enlarged and has an ameboid shape, the endoplasmic reticulum and other cytoplasmic organelles are increased, and wall ingrowths appear. These fingerlike projections into the cytoplasm are bordered by plasmalemma, thus increasing membrane surface available for moving materials into and out of the cell's interior. Transfer cells differentiate in a few days. However, their development can be blocked by shutting off the demand.

Syncytia induced by *Heterodera* and giant cells induced by *Meloidogyne* are examples of very large transfer cells. Wall ingrowths develop in 3 – 6 days in giant cells of *Impatiens balsamina* responding to *Meloidogyne*, and other morphological characteristics of transfer cells also appear. Thus a large part of the histopathology of certain nematode infections can be understood as the normal, nonpathological plant response to presence of a metabolic sink. When the sink is removed by death of the nematode, giant cells rapidly deteriorate and are replaced by undifferentiated parenchyma tissue. Differences in host reaction to each species or genus of nematode are superimposed upon the basic transfer cell response, so that the detailed pathology of *Heterodera* infections differs from that of *Rotylenchulus, Nacobbus*, and *Meloidogyne* even in the same host.

The spectacular cell alterations induced in hosts by sedentary endoparasites have intrigued investigators for many years. Apparently there are several components of the stimulus inserted into cells:

1. A message for enlargement of the cells upon which the nematode feeds and more generally in galls;
2. Another for blocking wall formation but without halting mitoses, as in the giant cells induced by *Meloidogyne*;
3. Another for partial dissolution of cell walls as in syncytia induced by *Heterodera*; and
4. A message for DNA replication in *Meloidogyne*-induced giant cells.

Minor changes in these components could result in a variety of cell responses. There is an interaction of parasite and host because the same host infected with two different endoparasites displays characteristic and different responses to each. The same nematode species induces giant cells that differ in detail in various hosts.

As studies of Heteroderidae and Meloidogynidae have ranged more widely, an array of variations has come to light. All species of *Heterodera* and *Globodera* induce syncytia consisting of partly fused cells without nuclear division; species of *Meloidogyne* induce giant cells; *Sarisodera* spp. (Heteroderidae) induce a single huge giant cell with a relatively enormous nucleus. This kind of nutritive cell also develops in infections with several other genera of Heteroderidae and Meloidogynidae, as well as with an unrelated nematode, *Rotylenchulus macrodoratus*.

Esophageal gland secretions are obvious sources of stimuli to plant cells, but they are not the only ones. Nematodes are metabolizing organisms that invade plant tissues. Consequently products of their metabolism—excretions from the excretory pore, feces, and possibly metabolites moving through the cuticle—are present in the immediate vicinity of nematodes. These contain ammonia, amino acids, and probably other organic compounds. Further, nematodes that deposit eggs within plant tissues are also releasing products from the gonads. *Meloidogyne* spp. often deposit egg sacs within plant tissues. These are secretions of rectal glands. And lastly, amphids release exudates. Until detailed studies of compounds from these sources are made, we cannot exclude products of nematode metabolism other than esophageal gland secretions as possible regulators of plant cell responses.

Damage to plants from nematodes is caused by cell destruction and by malfunction of any of the various systems that contribute to normal plant growth. Light to moderate numbers of parasites may not cause much injury but large numbers severely damage or kill hosts. The degree of damage depends on many factors. Under conditions of stress, such as drought, inadequate nutrition, or attack by other pathogens, nematodes add an additional stress, and the plant may die. In favorable circumstances a plant may survive the same intensity of attack that would kill it under less favorable conditions.

The complexity of disease is well illustrated by *Rotylenchulus reniformis* on sweet potato. In a variety trial, 'Goldrush' supported the least nematode reproduction but was most heavily damaged. Nematodes reproduced best on 'Centennial' but this variety was least damaged. We conclude that the biomass of nematodes is not sufficient to diminish growth of Centennial sweet potatoes but cell necrosis or other plant response in Goldrush interferes with growth (2). A similar effect occurs in certain plants resistant to *Meloidogyne*. Cell necrosis around invading juveniles damages roots and the nematodes die without inducing other pathologies. Treatment with nematicide reduces nematode populations and improves plant growth both in such resistant plants and in susceptible hosts.

We conclude this review of pathology with a reminder that nematode

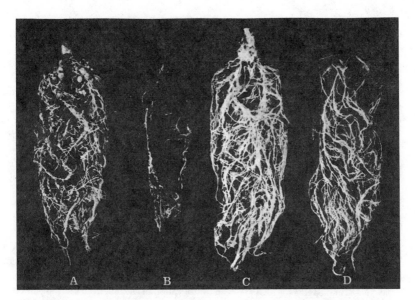

Fig. 7.2 Twelve-week-old tomato roots showing: (a) plants grown under aseptic conditions with *Meloidogyne incognita*; (b) plants grown under septic conditions with *M. incognita*; (c) and (d) plants grown under aseptic and septic conditions, respectively, but without *M. incognita*. (From P. S. Mayol, and G. B. Bergeson. 1970. The role of secondary invaders in *Meloidogyne incognita* infection. J. Nematol. 2:80 – 83, by permission.)

action enhances invasion of plants by bacteria and fungi. Plants in the field exist in an environment harboring pathogens of many kinds. Nematodes impose stresses on plants, and in many cases predispose their hosts to invasion by other pathogens. It is these interactions that make nematodes dangerous. Figure 7.2 illustrates the difference in root response to *Meloidogyne* when bacteria and fungi are absent and when these microorganisms are present.

REFERENCES

1. Barth, P., Stelzer, R., and Wyss, U. 1983. Changes in the inorganic metabolism in sugarbeet infected with *Heterodera schachtii*. Kali-Briefe 16:627 – 638.
2. Clark, C. A. and Wright, V. L. 1983. Effect and reproduction of *Rotylenchulus reniformis* on sweet potato selections. J. Nematol. 15:197 – 203.
3. Clausen, W. and Bird, A. F. 1984. The influence of root-knot nematodes (*Meloidogyne javanica*) on yield and on activity of sucrose synthase and invertase in roots of eggplants (*Solanum melongena*). Physiol. Plant Pathol. 25:209 – 217.

4. Glazer, I., Apelbaum, A. and Orion, D. 1985. Effect of inhibitors and stimulators of ethylene production on gall development in *Meloidogyne javanica*-infected tomato roots. J. Nematol. 17:145 – 149.

5. Huang, J. S., Barker, K. R. & Van Dyke, C. G. 1984. Suppression of binding between rhizobia and soybean roots by *Heterodera glycines*. Phytopathology 74:1381 – 1384.

6. Ko, M. P., Barker, K. R., and Huang, J. S. 1984. Nodulation of soybeans as affected by half-root infection with *Heterodera glycines*. J. Nematol. 16:97 – 105.

7. Ko, M. P., Huang, P., Huang, J. S., and Barker, K. R. 1985. Accumulation of phytoferritin and starch granules in developing nodules of soybean roots infected with *Heterodera glycines*. Phytopathology 75:159 – 164.

8. Krauthausen, H. J. and Wyss, U. 1982. Influence of the cyst nematode *Heterodera schachtii* on relative changes in the pattern of free amino acids at feeding sites. Physiol. Pl. Path. 21:425 – 436.

9. Nasr, T. A., Ibrahim, I. K. A., El-Azab, E. M., and Hassan, M. W. A. 1980. Effect of root-knot nematodes on the mineral, amino acid and carbohydrate concentrations of almond and peach rootstocks. Nematologica 26:133 – 138.

10. Poehling, H. M., Wyss, U., and Neuhoff, V. 1980. Microanalysis of free amino acids in the aseptic host-parasite system: *Ficus carica- Xiphinema index* (Nematoda). Physiol. Plant Pathol. 16:49 – 61.

11. Price, N. S. and Sanderson, J. 1984. The translocation of calcium from oat roots infected by the cereal cyst nematode *Heterodera avenae* (Woll). Rev. Nématol. 7:239 – 243.

12. Rinaudo, G. and Germani, G. 1981. Effect of the nematodes *Hirschmaniella oryzae* and *H. spinicaudata* on the N_2 fixation in the rice rhizosphere. Rev. Nématol. 4:171 – 172.

13. Viglierchio, D., 1971. Nematodes and other pathogens in auxin- related plant-growth disorders. Bot. Rev. 37:1 – 21.

14. Wang, E. L. H. and Bergeson, G. B. 1974. Biochemical changes in root exudate and xylem sap of tomato plants infected with *Meloidogyne incognita*. J. Nematol. 6:194 – 202.

15. Wyss, U. 1978. Root and cell response to feeding by *Xiphinema index*. Nematologica 24:159 – 166.

16. Wyss, U. and Zunke, U. 1986. Observations on the behavior of second stage juveniles of *Heterodera schachtii* inside host roots. Rev. de Nématol. 9:153 – 165.

17. Zacheo, G., Arrigoni-Liso, R., Bleve-Zacheo, T., Lamberti, F. and Arrigoni, O. 1983. Mitochondrial peroxidase and superoxide dismutase activities during the infection by *Meloidogyne incognita* of susceptible and resistant tomato plants. Nematolog. Mediterr. 11:107 – 114.

Eight

ECOLOGY

INTRODUCTION

Phytonematodes occur wherever there are plants. But the species of plants and of nematodes in Arctic tundra are not the same as those in tropical rainforests or in fields of corn. What controls the distribution of phytonematodes? What regulates numbers of nematodes at planting time in agricultural fields?

The relations between organisms and their living and nonliving environments are the subject of ecology. This branch of biology, derived from the study of natural history, has developed at two levels:

1. Adjustments of individual organisms to their environments— physiological responses to physical factors including light, moisture, temperature, and others; methods of obtaining nutrition and quantities consumed; patterns of reproduction; energy expenditures of various life stages; and general behavior.

2. Relations among the many forms of life living together. A major task is to understand how plants, animals, and microorganisms are integrated into communities, and why each habitat has its own array of species. Attention is therefore focused on food chains (who eats whom) and of fluctuations of populations of various species within communities. Factors affecting population growth, including immigration, reproductive rates, and predation are considered. Some ecologists construct mathematical models to predict population changes over time as conditions vary.

Most investigators concentrate on easily visible organisms in large areas, for example, forests, prairies, and intertidal zones. Ecological concepts, such as that of a climax forest, do not always apply on a microscale. Soil conditions vary greatly as we examine smaller and smaller environments. A nematode's habitat may consist of the space within a few millimeters of a growing root. Acquisition of meaningful data from soil takes relatively enormous amounts of labor; distribution of organisms is usually extremely patchy. Consequently, generalizations about soil ecology are risky (14).

Nematodes eat smaller organisms, graze on plant roots, and invade plants and animals, including humans. In turn, they are consumed by other forms of life. We compete with other organisms for energy. We manipulate various species, especially in agriculture. We alter the environment to suit ourselves. In so doing we may create favorable conditions for nematodes, which then become pests. Our task is to understand the ecology of nematodes sufficiently well to avoid problems of our own creation.

DISTRIBUTION

A. Variability. The distribution of nematodes in soil is highly variable, both in space and time. Phytonematodes cluster around growing roots and remain in the same vicinity when the plants die. In fields of row crops there are more nematodes *in* the rows than *between* rows. Moreover, along any single row, there are local regions of above-average concentrations. Plowing the soil before planting tends to increase uniformity of distribution, but in each season, variability of distribution in space is reestablished. Similar variations occur in time. Populations decline over winter in temperate climates and increase during the growing season. Nematodes have the reproductive capacity and short enough life cycles to build large populations during a single season; they are highly sensitive to variations in soil conditions and to availability of food. Differences in concentrations of nematodes develop during a single growing season (6, 11).

B. Passive Transport. Although nematodes move only a few centimeters from one location to another by their own efforts, they can survive during movement of soil and water. In dry periods wind carries nematodes, and in wet seasons floodwaters distribute them. Irrigation distributes nematodes when water from one field passes to the next. The Dutch have a long history of reclaiming agricultural lands from the sea. Crops first planted on these lands are free of phytonematodes. In a few years, however, damage by nematodes becomes apparent. Nematodes survive drying, and soil is easily carried about. We must assume that they are transported from one region to the next without difficulty.

C. Toposequences. Each species has its own tolerance limits and preferences for the various environmental components such as soil type, temperature, and humidity. Nematode species found in tropical soils are different from those in temperate regions, and those in regions of low rainfall differ from those in moist environments. A striking example of the close relation between environmental factors and species composition of soil fauna was described in Iowa (15). Norton and Oard sampled soil in a hilly, deep silty loam area where steep slopes prevail on which corn is planted on the contour. Plots, called toposequences, were established to determine nematode distribution at the tops, and descending along the slope to the level portion at the bottom (Fig. 8.1). Variations in pH, soil texture, organic matter, cation exchange capacity, and moisture were recorded. Both corn yields and nematode distributions varied along the slopes. "In general, a species tended to have a marked affinity for a par-

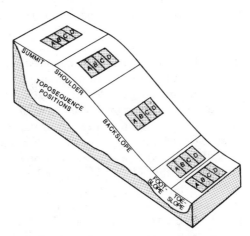

Fig. 8.1 Diagram of toposequence positions. (From D. C. Norton, and M. Oard. 1981. Plant-parasitic nematodes in loess toposequences planted with corn. J. Nematol. 13:314 – 321, by permission.)

ticular slope position." *Paratylenchus microdorus* was most numerous at the summit and shoulder; *Helicotylenchus pseudorobustus* peaked in the back slope; and *Pratylenchus* spp. was most numerous at the bottom of the slope. This careful study illustrates the importance of microenvironment in regulation of nematode populations. The authors also determined that the middle portion of north-facing slopes had greater diversity of species than the west-facing slopes because conditions were more uniform along the north-facing slopes. They suggested that locations with greater diversity of species have less potential for disease outbreaks because greater diversity indicates greater internal controls of nematode populations (15). However, this view is not shared by some contemporary ecologists.

D. Ecotypes. Regulation of populations by environmental factors applies not only to diverse species, but it also operates within a single species. Various populations are adapted to particular conditions within the broad range of a species. These subspecific variants are called **ecotypes**. *Heterodera avenae*, the cereal cyst nematode, provides an excellent example. A French investigator compared 24 populations from France, Belgium, and Switzerland. He identified four races, each with a different host range. There were two from France with different physiologies: Fr 1 from southern, and Fr 4 from northern localities. Fr 1 becomes active during the winter. Hatching begins at 5°C and is inhibited by temperatures above 10°C. Fr 4 becomes active in the spring; eggs hatch when soil warms to 10°C or higher. These differences persisted when both populations were maintained for 3 years in an intermediate locality, and subsequently tested in all three places (16). Ecotypes of other nematode species have also been reported.

E. Marine Nematodes. Although marine nematodes are not phytonematodes, a brief mention of this group is appropriate. "Nematodes are the most abundant metazoans in marine sediments, extending from the high-water mark into the deepest oceanic trenches. . . . In most circumstances, nematodes are numerically the dominant group of the meiofauna (small, but not microscopic organisms). In fact, they usually comprise more than 90% of the metazoan fauna" (8). Species composition in different habitats is distinctive; salinity, character of the sediment, and oxygen concentration are important ecological factors. In general these nematodes are larger than soil nematodes and many of them have setae and distinctive amphids. Most marine species are classified into orders not found in soil. Large populations have been recorded in salt marshes with a high mud content ($23,000,000/m^2$). Most of the population occurs in the upper few centimeters of sediment, but some nematodes reach lower, completely anoxic layers. Food of marine nematodes consists of bacteria, algae, and each other. The algae make their own food by photosynthesis and bacteria live on debris that enters the system from other sources. Investigators are attempting to characterize changes in species composition as an index of pollution, but results are not consistent.

F. Population Densities. Soil contains fewer nematodes than marine habitats. Published estimates of total nematodes/m² in 81 sites representing eight different ecosystems show average values ranging from a low of 760,000 nematodes/m² in deserts to a high of 9,190,000 in grasslands. Tropical forests have lower numbers of nematodes (1,700,000) than broad-leaf forests of Europe (6,270,000). Coniferous forests have about half as many nematodes as deciduous forests (20).

G. Soil Texture. Soil properties are important for nematodes. Nematodes build large populations in sandy soils; less frequently there is a preference for clay soils or heavy, wet soils; and in a few cases soil texture has no influence. Some investigators report that a nematode species prefers light soils, and others find that it prefers heavy soils. More than texture is involved in the regulation of population growth. Texture is related to pore size distribution and to the behavior of water in soil. It is one factor among many that determine suitability of soils for nematodes. Texture is important in pathogenicity because it affects the rate of drying. Lower concentrations of nematodes cause more damage in a soil with poor water-holding capacity than in one with good to excellent water-holding capacity.

Seinhorst's description of onion bloat in Holland illustrates the influence of soil texture. He estimated populations of *Ditylenchus dipsaci* in agricultural soils of the Netherlands by repeated samplings of many accurately located sites. This species occurs in all Dutch river marine clay soils that he sampled. In an intensive study of an island in the Rhine delta, Seinhorst demonstrated a clear relation between clay content and intensity of disease (17). "Onion bloat is a persistent menace on all heavy clay soils whereas on light soils it only becomes important when onions are grown . . . more than once in three or four years." Figure 8.2 is a comparison of a soil map and a map of the distribution of damaged onions. "There is a great degree of coincidence between the area nonsuitable for onions because of the persistent menace of bloat and the area with clay content higher than 30%." Nematode populations decreased more rapidly on sandy and light soils than on heavy soils in the absence of host crops during winter. Populations increased to high levels on onions, rye, and oats in sandy soils, but fell to low levels over winter. Seinhorst suggested that weeds maintain populations at approximately 10/100 g of soil in heavy clays. As few as two to four

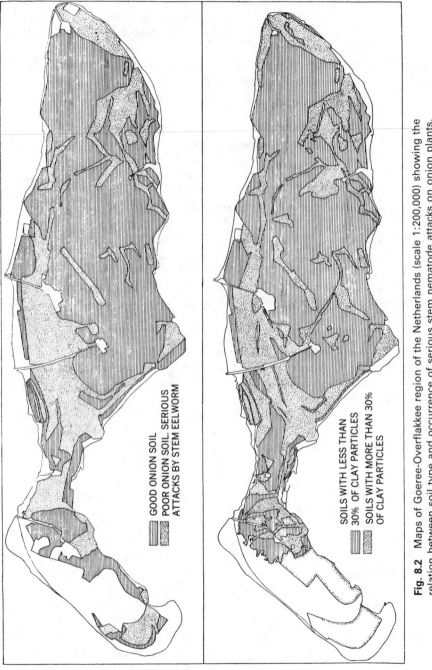

Fig. 8.2 Maps of Goeree-Overflakkee region of the Netherlands (scale 1:200,000) showing the relation between soil type and occurrence of serious stem nematode attacks on onion plants. (From J. W. Seinhorst. 1956. Population studies on stem eelworms (*Ditylenchus dipsaci*). Nematologica 1:159 – 164, by permission of E. J. Brill.)

GOOD ONION SOIL

POOR ONION SOIL. SERIOUS ATTACKS BY STEM EELWORM

SOILS WITH LESS THAN 30% OF CLAY PARTICLES

SOILS WITH MORE THAN 30% OF CLAY PARTICLES

individual *D. dipsaci*/100 g may cause severe onion bloat and when there are 5/100 g of soil, no healthy plants can be found in the crop.

Additional evidence of the influence of soil conditions on nematodes is reported in Ref. (22). Townshend packed polystyrene cylinders with soil of various textures. A series of combinations included a sandy loam, a silt loam, and a loam soil, two bulk densities, six moisture tensions, three temperatures, and two nematodes, *Pratylenchus minyus* and *P. penetrans*. Optimal temperature for penetration into corn roots was 20°C for *P. penetrans* and 30°C for *P. minyus*. Penetration of both species was most successful in all soils at moisture tensions between 10 and 100 cm of water and at low bulk densities. More nematodes entered corn roots in the sandy loam than in the other two fine-textured soils. Both of these examples show that nematodes do not change soil conditions to suit themselves, but operate within limits set by the environment in their immediate vicinity.

POPULATION DYNAMICS

Sampling

Every investigator of populations, whether of nematodes or of other organisms, will ask: How accurate is my estimation of numbers of organisms present? Can I count every one or must I count samples of the population? Where is the population? When does it peak and when does it decline?

Errors are inherent in any sampling procedure to estimate populations of nematodes in soil. The process may be divided into the following steps, each subject to error:

1. Collection of soil cores from field, garden, or orchard;
2. Transport of samples to laboratory and storage before processing;
3. Extraction from soil;
4. Counting nematodes;
5. Statistical analysis of data.

Nematologists have designed many sampling schemes to minimize labor and yet provide adequate estimates of numbers and kinds of nematodes present (1, 4, 6, 12). Soil cores are usually obtained from the upper soil (15 – 20 cm) where roots are present, as in the drip line of trees, and in the rows of crops, as well as from areas between these. If the nematodes prefer greater depths, as in some distributions of *Trichodorus* and *Longidorus*, cores down to 60 cm or more may be appropriate. Cores are combined, the soil is mixed, and subsamples are taken for extraction and counting. One hopes to get within 20 to 25% of the "true mean" of the population. As McSorley and Parrado (10) state in their conclusion, "It is apparent that sampling plans must be custom made for many different situations, since the relationship between number of cores and relative error changes in response to many factors, including nematode species and density, field size, crop and soil type." Figure 8.3 illustrates a sampling scheme together with an estimate of its precision.

Population analyses are often made by a few technicians, who are usually under pressure of other work. Samples must be collected in the field, chilled, then brought to the laboratory and protected during storage until counts can be made. Many nematodes die during transport and storage, especially in warm weather.

All techniques of extraction from soil yield only a fraction of the nematodes present. To analyze each fragment of soil is obviously not practical. Soils differ, and the technique suitable for a light soil may not be suitable for a heavy soil. Those with much organic matter are particularly difficult. It is important, therefore, to estimate the proportion of nematodes found by the particular technique used. Townshend tested a time-honored method, the Cobb decanting and sieving procedure, by adding known numbers of nematodes to soil, then extracting them. Soil texture, size of sample, density of population, and species of nematode all influenced efficiency of extraction. "Approximately 25% of a population of *Pratylenchus*

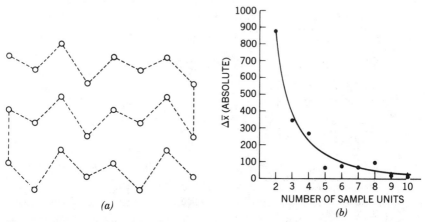

(a) *(b)*

Fig. 8.3 (a) Pattern for collecting soil cores in a fallow field or established row or field crop; (b) relationship between precision of population assessment and number of sample units. (From H. Ferris. 1985. Sampling, precision and reliability. In: Plant Nematology Laboratory Manual, (B. M. Zuckerman, W. F. Mai, and M. B. Harrison, eds.). Univ. Massachusetts Agric. Exp. Stn, Amherst, pp. 21 – 25, by permission.)

penetrans could be extracted from 454 g samples of fine sandy loam containing 3000 or more nematodes. Below this density, yields dropped sharply" (21).

Statistical analysis of counts is currently receiving increasing attention. Wallace (23) has an excellent discussion of the distribution and abundance of nematodes (pp. 86 – 115). He points out that counts usually have a greater variance than their means, a sign of patchy distribution.* Because nematodes are clustered in soil, a plot of distribution of counts is never the bell-shaped curve of normal distribution. The frequency distribution of counts is always skewed to the left, with low counts more numerous than high. The negative binomial and other distributions are used to facilitate statistical analysis of data (3, 9). Figure 8.4 shows the effect of number of samples on accuracy of the data.

Seasonal Cycles

Conditions in soil change as the seasons advance. Moisture content, temperature, abundance of growing roots—all proceed in their cycles through the year. Nematode populations also respond to changing conditions.

A. *Meloidogyne* and *Xiphinema* in a Vineyard: Distribution of populations of *Meloidogyne* spp. and *Xiphinema americanum* has been mapped in soil of an irrigated California vineyard throughout the year (5). Eggs and juveniles of *Meloidogyne* are at a low level in summer until late August when eggs accumulate rapidly. At this time the soil is dry because rain is scanty and irrigation is normally withheld to permit grapes to increase their sugar content. Eggs do not hatch until irrigation is applied in September after harvest and the number of infective juveniles increases. Both eggs and juveniles survive winter. Soil warms up in spring and a new group of eggs is produced, leading again to a subsequent increase of juveniles which then invade emerging rootlets in April and May. Throughout summer, soil populations remain at a low level.

Xiphinema americanum populations increase dramatically in autumn, remain high over winter, and decline during February to a steady low level in spring and summer. Although the life cycle of *Xiphinema* in California vineyards is not precisely known, we surmise that this ectoparasitic nematode depends more on soil conditions than the endoparasitic *Meloidogyne*. Moreover, a *Meloidogyne* female produces many eggs that remain quiescent until conditions are favorable for hatching. Juveniles then enter

---

* $\Sigma (x - \bar{x})^2/n - 1 > \Sigma x/n$, where x = any count, $\bar{x}$ = mean, n = number of counts, Σ = sum.

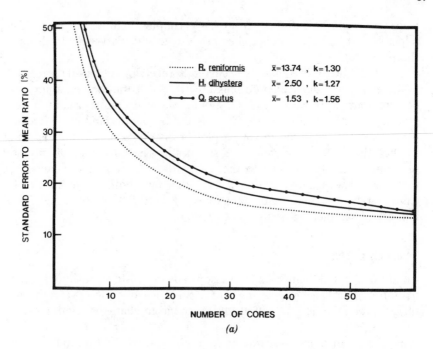

(a)

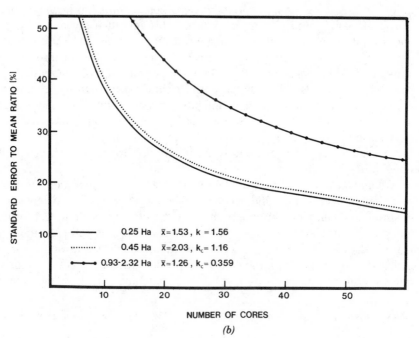

(b)

plants quickly. A *Xiphinema* female, on the other hand, spends much more of its life cycle in soil, produces relatively few eggs, and is believed to be long-lived.

Spatial distribution of each nematode population in the vineyard is distinctive. *Meloidogyne* nematodes are most concentrated in the upper 60 cm of soil; most of them are close to the vines. Some individuals occur between rows, especially at the center where roots from adjacent rows overlap. There are appreciable numbers of nematodes down to 120 cm. *Xiphinema* are concentrated in the top 45 cm in the vine rows and do not occur at all frequently between rows. Vines are maintained in a ridge of undisturbed soil. Cultivation of the region between rows is by disking and furrowing at intervals. The authors of this study suggest that *Xiphinema americanum* nematodes are confined to upper soil layers because they are sensitive to low oxygen tensions in deeper layers. Machine operations between rows of vines probably prevent this nematode from increasing its population away from undisturbed ridges where the vines are.

B. *Meloidogyne* on Cotton: Figure 8.5 illustrates seasonal cycles of three species of phytonematodes feeding on five kinds of cotton. Population size and timing of the cycles differ among the species. Many factors interact to influence reproductive success of phytonematodes, and this diagram shows the effects of genetic resistance. *Meloidogyne* reproduces better on Rowden cotton than on *Gossypium barbadense* or on Auburn 56. Fewer nematodes are present on resistant cottons throughout most of the growing season; populations peak only when the plants finish most of their growth. Each geographic region and each local site has its own characteristic population cycles. It is important to know these for design of management practices.

Population Models

Models of nematode populations and plant yields are mathematical statements and computer programs representing portions of presumed reality. The maker's goal is to predict populations and crop yields on the basis of limited information about a particular situation. For example, if you know

---

Fig. 8.4   Relationship between number of cores per sample and size of plot sampled to relative error. (a) Relative error in terms of standard error to mean ratio for three nematode species in a 0.25-ha plot. $\bar{x}$ = mean of all samples from a given field; k = value from negative binomial distribution. (b) Relative error as in (a) for *Quinsulcius acutus* (similar to *Tylenchorhynchus* sp.) in plots of various sizes. (From R. McSorley, and J. L. Parrado. 1982. Estimating relative error in nematode numbers from single soil samples composed of multiple cores. J. Nematol. 14:522 – 529, by permission.)

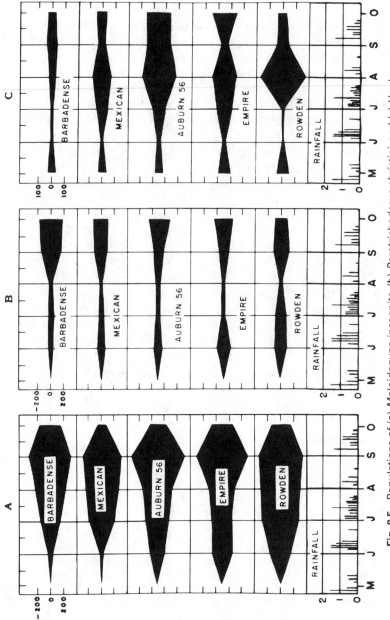

Fig. 8.5 Populations of **(a)** *Meloidogyne incognita*, **(b)** *Paratrichodorus christiei*, and **(c)** *Hoplolaimus tylenchiformis* in cotton plots. (From D. C. Norton 1978. Ecology of plant parasitic nematodes, 268 pp. Wiley, New York, by permission of John Wiley & Sons.)

the initial nematode population when a crop is planted, can you predict the final population at the end of a growing season, and that at the start of next year? It may seem hopeless in the face of the known number of variables that affect nematodes and hosts, but progress is being made.

Every model builder uses information from the past to predict the future. If the programs accurately portray nematode population cycles, the model's predictions will enable a grower to apply control measures at the best time. In fact, it is hard to build an accurate model. Success requires thorough knowledge of each component in the web of interrelationships existing in the field. Every nematode population has its own physiological properties. Some species reproduce with short life cycles and display great fecundity whereas others have longer life cycles or produce fewer offspring. Every crop has its own suitability as host to phytonematodes. Each field has its own capacity to foster or hinder nematode reproduction, and this capacity will affect some species differently from others. In addition, the presence of populations of organisms other than the one of interest may affect nematodes. A model builder soon realizes that the base of quantitative knowledge is limited and that a major task is to acquire better knowledge. As part of this effort one may insert different values for each environmental component, in turn, into the model and thus compare the relative influence of each component on population dynamics.

Early efforts described population changes and yield reductions from field data. A graph of plant growth or yield plotted against log nematode density is closer to a straight line than a plot of growth or yield against actual numbers of nematodes. This seems logical because the addition of a small increment to a scant population should have more effect than the addition of the same increment to a large population.

Seinhorst (18,19) produced models based on theoretical considerations, then tested them in a series of careful experiments. He was the first nematologist to propose a model based on general biology of nematode – plant associations. He believed that in predicting yields of crops in the presence of nematodes, one must collect data from a wide range of samples including low to high nematode densities. He noted that each plant – nematode association has a characteristic level of nematode population at which damage is apparent. He also realized that plants produce more roots than the amount required for good top growth. His model is based on the competition curve of Nicholson (11). Seinhorst assumed that the "average nematode" does not vary as population density changes and that the ability of nematodes to cause damage does not change at high densities. His final formula, produced after years of work and thought, is

$$Y = m + (1 - m)Z^{P-T}$$

where Y = yield; m = relative minimal yield when nematode density is at its maximum; Z is a constant < 1, representing the fraction of roots not attacked by nematodes; P = nematode population density; and T = tolerance level.*

To state this in words, the yield of a crop under attack by nematodes is the sum of the low yield when nematodes are at the maximum level known for that crop plus an increment that depends on the particular nematode population observed. This increment $(1 - m)$ is affected by the proportion of roots that escape infection. This proportion, Z, is raised to an exponent. The exponent consists of the difference between the observed density of nematodes and that at which no damage results $(P - T)$. Under carefully controlled conditions, Seinhorst varied the number of nematodes and examined growth and yield. He fitted curves derived by the model to observed data and determined values of Z and T that produced the closest fit. This model is certainly not perfect, but it represents a major accomplishment and has stimulated others to build their own models.

The arrival of computers made it practical to develop mathematical models to predict population behavior in the field. Models are being developed with computers that permit an investigator to test predictions under simulated conditions. Recent examples are given to illustrate different approaches to modeling. Noling and Ferris (13) explore the "ceiling" reached by nematode populations on hosts. Each habitat is presumed to have its own "carrying capacity" for particular organisms. That is, the environment can support a certain size of population, and when this is reached, the population remains more or less constant at that level.

The authors tested the existence of a ceiling population for *Meloidogyne hapla* on alfalfa growing in microplots out of doors. They inoculated various numbers of nematodes ranging from 0 to 2170/1000 cm³ of soil in the microplots at the same time as they planted alfalfa. They removed plants at intervals and counted nematodes in alfalfa roots as well as in soil. Measurements of many kinds were taken: root weight, eggs and juveniles per gram of root, rate of population change, numbers of mature females per

---

* The relationship between number of nematodes and proportion of undamaged root tissue. (Reproduced from Ref. (7), p. 32.)

Number of Nematodes	Proportion of Root Damaged	Proportion of Root not Damaged
1	d	$(1 - d) = z$
2	$d + d(1 - d)$	$(1 - d)^2$
3	$d + d(1 - d) + d(1 - d)^2$	$(1 - d)^3$
P		$(1 - d)^P = z^P$

gram of root, and average numbers of eggs/female. They all showed that as the population approached a ceiling level, reproduction slowed until it reached a rate that just replaced the population. Of course, in addition to effects of crowding and limitation of root supply, other factors act in density-dependent ways. Predators and parasites tend to increase as nematode populations rise. Other organisms that feed on roots will depress nematodes. And catastrophes for plants and nematodes, such as abnormal temperatures, flooding, or strong windstorms may intervene. The data in this report are graphed against accumulated heat units (degree-days above 10°C) and the model appears as a regression of the "relationship between population $\log_{10}$ multiplication factor (P_f/P_i) and $\log_{10}$ initial population density (P_i)." P_f refers to the final population. As the initial population increased, the rate of multiplication declined.

The second paper describes an attempt to develop a simulation model of *Heterodera schachtii* infecting sugarbeet (2). Caswell et al. prepared a computer program to calculate increase of nematode numbers and growth of beets at 72 min intervals. A number of simplifying assumptions were made and data were taken from published literature. "The model simulates population dynamics of the nematode on a single sugarbeet plant in a soil volume approximating the volume a single plant occupies in the field." Figure 8.6 presents flow charts of the computer program.

Models such as this iterate calculations of changes in short time intervals using temperature as the driving factor. Comparison of computer results with field data measures the accuracy of the model. The major problem is lack of accurate data. Perhaps the greatest result of current efforts will be that nematologists will begin to collect useful data upon which to build good mathematical and computer statements of population dynamics and yield reductions.

Community Structure

How important are nematodes for the life of the soil? How much of the energy flow through soil passes through nematodes? Are the organisms living within soil and on its surface integrated in any way into a community that returns to a more or less stable condition after it is perturbed? Nematologists are beginning to address such questions.

Through photosynthesis and subsequent metabolism, plants assemble complex, energy-rich molecules: the carbohydrates, proteins, lipids, and other components of plant structure. Phytonematodes tap these stores of energy by removing cell contents through their stylets. Other nematodes feed on bacteria and fungi that degrade plant and animal remains in soil. And in turn, living nematodes are hosts of bacterial and fungal parasites

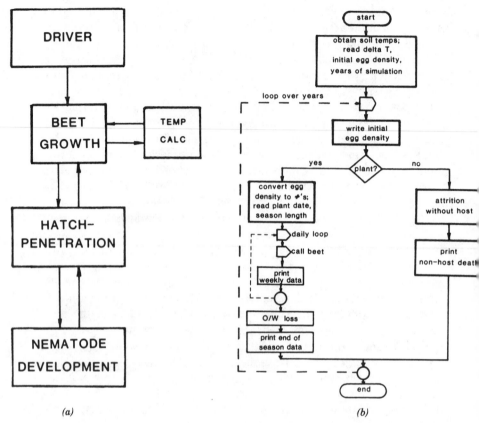

Fig. 8.6 Flow chart of computer simulation of *Heterodera schachtii* populations on sugar beet. (**a**) Logical flow of the model, indicating interactions among subroutines. (**b**) flow chart of the main program. (From E. P. Caswell, M. K. MacGuidwin, C. E. Nelsen, I. J. Thomason, and G. W. Bird. 1986. A simulation model of *Heterodera schachtii* infecting *Beta vulgaris*. J. Nematol. 18:512 – 519, by permission.)

and are consumed by a variety of predators including earthworms, carnivorous nematodes, small insects and other arthropods, mites, and protozoa. Remains of nematodes are disintegrated by bacteria and fungi.

Bacteria, fungi, and actinomycetes, as well as plants, are the principal agents of soil metabolism. Quantities of energy consumed by nematodes can be estimated by determining their biomass and rate of energy exchange per nematode. It is impractical to isolate all nematodes in soil and to weigh them, but volume and dry weight can be calculated from measurements of body size and water content. Respiration of individual nematodes can be measured with sophisticated instruments. By these methods, a few investigators have attempted to measure energy flux through nematodes in soil. In general, nematodes represent a small proportion of the total biomass. However, they may influence the activity of other organisms to a considerable extent. Estimates of biomass from 49 sites representing seven different ecosystems ranged from 410 to 3800 mg/m^2 (20). Individual nematodes were calculated to have fresh weights from 0.2 to 0.5 μg. The author concluded that "nematodes could be responsible for 10 to 15% of the soil animal respiration."

Another estimate offers different values. Phytonematodes constitute only a fraction of the total nematode fauna, from 20 to 33% in data from Polish potato fields. The total soil fauna is responsible for 5 – 10% of the heterotrophic respiration. Hence nematodes account for approximately 1% of soil metabolic activity or less.

Some ecologists are taking a new approach to the complexity of soil – nematode relationships by creating miniature environments, "microcosms," in which they measure the role of a few members of the soil biota separately and in combinations (7).

Application of pesticides often has unexpected results because food chains in soil are usually complex, involving many different species. Most nematodes in soil feed on bacteria and fungi. They may therefore lower the rate of decomposition of organic matter by depressing populations of decomposers. In turn, nematode populations are subject to predation, especially by mites. Pesticides may remove some but not all species of the soil biota, thus permitting remaining species to multiply vigorously. *Trichodorus* spp. often reach more dangerous population levels after soil fumigation than before treatment.

The role of nematodes in agricultural soils with conventional farming differs from that in other ecological systems. There are fewer species of organisms under crops, soil is periodically disturbed, and phytonematodes are often a greater proportion of the total nematode fauna under growing crops than in other ecosystems. Most organisms in soil find food to be scarce. When organic matter appears, food chains are activated around it, and populations of many organisms increase rapidly. Intense activity takes place in root rhizospheres. Organic compounds leak from roots, cells are sloughed off from root tips, and organisms that feed directly on roots are attracted to them. When roots die, the remains are decomposed by bacteria and fungi, and nematodes, as well as other "grazers," consume these microorganisms.

Conclusion

Many species, in addition to humans, feed on agricultural crops. Bacteria, fungi, mites, insects, birds, mammals, and nematodes share the wealth provided by human effort. Growers in highly technical agricultures devote substantial energy to suppression of pests. But growers in subsistence agriculture lack resources to do this. They must, therefore, rely on experience to achieve a harvest despite competitors. Until recently nematodes were unknown adversaries.

Modern ecology teaches us that life is a web of interacting species. Each species represents a unique form of life designed to gain access to as much energy as possible and to propagate its own kind. In our highly developed agriculture humans have disregarded much of the life in soil. As a result, productive land became unproductive, often ending as desert. We must now view nematodes as part of the web of life so as to accommodate to their presence and to produce our desired food and fiber while at the same time avoiding conditions leading to damaging populations of pests. Although nematodes usually affect underground parts of plants and do not cause major changes in the ecology of a region, an exception is the pine wilt nematode. This pest has killed a large proportion of susceptible pines in much of Japan, and soil erosion on steep slopes is now a serious problem.

Chapter 9 presents genetic resistance to nematodes and Chapter 10 is an overview of nematode control. Both subjects are rapidly evolving toward more sophisticated ways to satisfy the requirements of sustainable agriculture and to maximize yields while at the same time keeping ecological disturbances at a minimum.

REFERENCES

1. Barker, K. R. 1985. Soil sampling methods and procedures for field diagnosis. In: Plant Nematology Laboratory Manual, (B. M. Zuckerman, W. F. Mai, and M. B. Harrison, eds.). Univ. Massachusetts Agric. Exp. Stn, Amherst, pp. 11 – 20.

2. Caswell, E. P., MacGuidwin, M. K., Nelsen, C. E., Thomason, I. J., and Bird, G. W. 1986. A simulation model of *Heterodera schachtii* infecting *Beta vulgaris*. J. Nematol. 18:512 – 519.

3. Davis, R. M. 1984. Distribution of *Tylenchulus semipenetrans* in a Texas grapefruit orchard. J. Nematol. 16:313 – 317.

4. Ferris, H. 1985. Sampling, precision and reliability. In: Plant Nematology Laboratory Manual, (B. M. Zuckerman, W. F. Mai, and M. B. Harrison, eds.). Univ. Massachusetts Agric. Exp. Stn, Amherst, pp. 21 – 25.

5. Ferris, H., and McKenry, M. V. 1974. Seasonal fluctuations in the spatial distribution of nematode populations in a California vineyard. J. Nematol. 6:203 – 210.

6. Francl, L. 1986. Spatial analysis of *Heterodera glycines* populations in field plots. J.Nematol. 18:183 – 189.

7. Freckman, D., 1982. Nematodes in Soil Ecosystems, Univ. Texas Press, Austin.

8. Heip, C., Vincx, M., Smolm, N. and Vranken, G. 1982. The systematics and ecology of free-living marine nematodes. Helminthological Abstracts Series B, Plant Nematology. 51 (1): 1 – 31.

9. McSorley, R. 1982. Simulated sampling strategies for nematodes distributed according to a negative binomial model. J. Nematol. 14:517 – 522.

10. McSorley, R. and Parrado, J. L. 1982. Estimating relative error in nematode numbers from single soil samples composed of multiple cores. J. Nematol. 14:522 – 529.

11. Nicholson, A.J. 1933. The balance of animal populations. J. Anim. Ecol. 2:132.

12. Noe, J. P. and Campbell, C. L. 1985. Spatial pattern analysis of plant-parasitic nematodes. J. Nematol. 17:86 – 93.

13. Noling, J. W. and Ferris, H. 1986. Influence of Alfalfa plant growth on the multiplication rates and ceiling population density of *Meloidogyne hapla*. J. Nematol. 18:505 – 511.

14. Norton, D. C. 1978. Ecology of plant parasitic nematodes. Wiley (Interscience), New York, 268 pp.

15. Norton, D. C. and Oard, M. 1981. Plant-parasitic nematodes in loess toposequences planted with corn. J. Nematol. 13:314 – 321.

16. Rivoal, R. 1978. Biologie d'*Heterodera avenae* Wollenweber en France. 1. Differences dans les cycles d'eclosion et de développement des deux races Fr1 et Fr4. Rev. Nématol. 1:171 – 179.

17. Seinhorst, J. W. 1956. Population studies on stem eelworms (*Ditylenchus dipsaci*). Nematologica 1:159 – 164.

18. Seinhorst, J. W. 1965. The relation between nematode density and damage to plants. Nematologica 11:137 – 154.

19. Seinhorst, J. W. 1973. The relation between nematode distribution in a field and loss in yield at different average nematode densities. Nematologica 19:421 – 427.

20. Sohlenius, B. 1980. Abundance, biomass and contribution to energy flow by soil nematodes in terrestrial ecosystems. Oikos 34:186 – 194.

21. Townshend, J. L. 1962. An examination of the efficiency of the Cobb decanting and sieving method. Nematologica 8:293 – 300.

22. Townshend, J. L. 1972. Influence of edaphic factors on penetration of corn roots by *Pratylenchus penetrans* and *P. minyus* in three Ontario soils. Nematologica 18:201 – 212.

23. Wallace, H. R. 1973. Nematode ecology and plant disease. Arnold, London, 228 pp.

Nine

RESISTANCE

INTRODUCTION

A host and its parasite are like two dancers who respond to every subtle movement of each other as they weave a pattern of movement before us. Plants and nematodes have lived together for millions of years and have adjusted to each other in ways that we are just beginning to perceive. Results of coevolution range from the simple association of ectoparasites such as *Trichodorus* spp. with roots of corn to the complex life cycle of *Heterodera schachtii* in beet roots. The stubby root nematode grazes along outer root surfaces, feeding briefly on epidermal cells. Susceptible hosts support abundant reproduction and may be severely damaged by large parasite populations. Resistant hosts are less favorable and escape damage. Cyst nematodes, however, have a highly complex association with plants. Juveniles enclosed within eggs contained in the female's body are stimulated to hatch by compounds leached out of growing roots. The young nematodes are attracted to roots, penetrate, find places to settle, and stimulate host cells to form specialized feeding sites on which the parasites subsist. Both the form and metabolism of cells at feeding sites are altered in response to the parasites. Genetic controls in both organisms probably operate at each step of this process.

TERMINOLOGY

The terms we apply to hosts in the presence of nematodes reflect characteristics of both organisms. **Resistant** plants depress nematode reproduction and remain healthy. **Intolerant** plants also fail to support nematode reproduction but suffer damage. Nematodes multiply well in **susceptible** hosts and inflict damage. They also multiply well in **tolerant** plants, but inflict little damage.

We may visualize these relationships in a 2 × 2 diagram showing nematode health (= reproduction) on the horizontal side and plant health (= growth) on the vertical. + indicates good plant growth and good nematode reproduction; – indicates poor plant growth and poor nematode reproduction. The terms describe plants in the presence of phytonematodes.

Nematode Reproduction

		+	–
Plant Growth	+	Tolerant	Resistant
		PLANTS	
	–	Susceptible	Intolerant

Plant pathologists also use the term **virulent** parasites — those that thrive in particular hosts, and **avirulent** ones which do not invade or reproduce well in certain host – parasite combinations.

In all discussions of host – parasite relations we must remember that particular nematode populations and particular plant hosts are specified. Members of one species of nematode do not all have the same genetics. Genes vary from one individual to the next. Plants of a species also have variable genetics. Breeders select and cross many plants to produce uniform cultivars with desired qualities. However, cultivars differ from each other. Some may be susceptible and others may be resistant to certain nematodes. The term **nematode race** refers to a subdivision of a species with a distinctive host range. Nematologists in Europe generally use the word **pathotype** to refer to a population of phytonematodes distinguished from others by its ability to reproduce on some members of a set of hosts used as a standard test. Pathotypes of potato cyst nematodes are distinguished by their success or failure to reproduce on a set of *Solanum* clones used as sources of potato resistance to cyst nematodes.

The terms mentioned in this discussion must be viewed as attempts to organize information and not as designations of absolutes. Thus resistance, susceptibility, tolerance and intolerance all grade from high to low. There are differences in virulence; races (pathotypes) are often hard to separate. Moreover, environmental conditions may affect a plant- nematode association. For example, a gene that confers resistance of tomato to *Meloidogyne* spp. loses its effect at high soil temperatures.

ASSAYS FOR RESISTANCE

To find sources of resistance useful to plant breeders, nematologists search for resistant individuals in collections of seeds from species related to crop plants. These collections usually represent different geographical areas. Seeds are planted in soil infested with relevant nematodes; resistant or tolerant individuals reveal themselves. The breeder then crosses these with plants of good agronomic quality and tracks resistant plants through successive generations until a desirable, high-yielding, and vigorous group of offspring is obtained. Before seed can be released for commercial use, potential cultivars are tested in various environmental conditions to compare their potential both with and without nematodes. Breeders grow thousands of plants and make repeated crosses to incorporate the gene(s) for resistance in a useful genetic background. Often resistance to more than one pathogen is combined in a single cultivar. It may take 10 years of work to produce a successful cultivar resistant to certain nematodes.

Heterodera and *Globodera*

Beets and potatoes to be screened for resistance to cyst nematodes are usually grown in small pots of infested soil. Roots are examined at the end of a single life cycle for the presence of newly formed cysts. Each pot is inverted, and the soil ball removed by tapping. The observer estimates numbers of new cysts visible on the network of roots at the soil ball surface. Resistant and susceptible plants are easily distinguished. Oats, barley, and wheat plants are easily managed in test tubes or out of doors in clay tiles plunged into soil. Again, the object is to count cysts produced in one or more life cycles. Investigators have repeatedly tried to reduce labor and space required to assay young plants. Cylinders of aluminum foil filled with soil and maintained in growth chambers have been used. Eggs and juveniles obtained from crushed mature cysts may replace infested field soil to obtain uniformity of exposure to nematodes (cf Chapter 4, mechanical maceration, p. 58). Candidate plants may also be set out in infested fields. Known susceptible seedlings may be alternated with candidates to find resistance in soil with nonuniform intensities of inoculum. Susceptible plants indicate the intensity of infestation at each location in the field.

The assumption in such work is that seedlings express genetic resistance reliably. This seems to be the case in most associations of nematodes and plants, although some exceptions are known. Roots excised from seedlings and grown in culture are good indicators of resistance and susceptibility (15). In addition, seedlings with parts of the tops removed have reduced root systems that display the same host – parasite relations as intact plants. Such pruned seedlings are convenient to grow in small containers or in hydroponics (10).

Ditylenchus dipsaci

Ditylenchus dipsaci attacks cereals and pasture crops. Nematodes grown on callus cultures are widely used in assays for resistance. Nematodes are maintained on callus in test tubes with nutrient agar slants; subcultures are routinely made at intervals. This technique permits nematodes from different populations to be kept conveniently for use in selecting for resistance. Nematodes are quickly collected and aliquoted into uniform portions for inoculation. Seedlings grown in rolls of filter paper are held in small containers and are inoculated with sprays of nematodes suspended in water. Drops of water carrying known numbers of nematodes can also be used. Susceptible plants become dwarfed or galled. Resistant seedlings can be moved to soil in a greenhouse and subsequently tested in infested plots in the field (4). Figure 9.1 shows the inoculation and selection procedure in a plant breeding laboratory in Sweden.

Meloidogyne

Resistance to root-knot nematodes shows as diminished numbers of galls, although this is not true in every case. In a survey of soybean lines inoculated with several species of *Meloidogyne*, roots in one combination had conspicuous galls containing immature females without eggs, whereas susceptible plants had galls with large protruding egg sacs (6). Therefore, judgement based on number of galls would have overlooked this example of resistance. Some assays are based on egg counts or counts of juveniles (5).

Inoculum to identify resistance may consist of chopped roots, egg sacs, free eggs, or hatched infective juveniles. Roots of infected susceptible hosts bearing numerous egg sacs should be lifted and washed free of soil before they have rotted. They can be mixed with sterilized soil or placed in funnels under mist for collection of newly hatched infective juveniles. Eggs can be liberated from the gelatinous matrix of egg sacs with the aid of NaOCl (see Chapter 4, p.). A convenient system consists of a bed of sterilized soil in the greenhouse with rows of holes to receive seedlings together with water suspensions of eggs or hatched juveniles. Automatic repeating syringes or similar devices can be used to deliver uniform inoculum.

Fig. 9.1 Assay for resistance based on callus cultures. (**a**) Interior of incubator for nematode and callus cultures; (**b**) Equipment for infecting callus cultures with nematodes; (**c**) Inoculated red clover seedlings ready for selection. (1) stunted plants, (2) swollen plants, (3) healthy plants. (From S. Bingefors and K. Bengt Eriksson. 1968. Some problems connected with resistance breeding against stem nematodes in Sweden. Z. Pflanzenzuecht. 59:359 – 375, (by permission of Verlag Paul Parey, Berlin.)

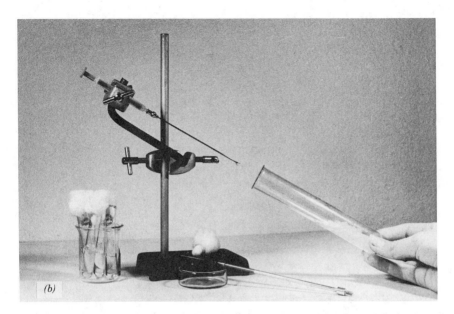

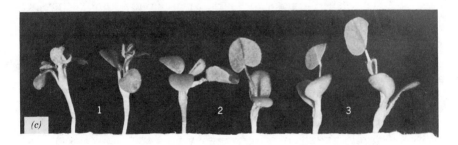

Fig. 9.1 (*Continued*)

Other Nematodes

All assays of resistance require large quantities of inoculum. Many nematode species can be grown on callus cultures of a variety of plants. This technique is convenient and economical of space and labor. A few species of phytonematodes such as *Bursaphelenchus xylophilus*, some species of *Aphelenchoides*, and *Ditylenchus destructor* reproduce on fungi, offering convenient materials for producing inoculum.

MODES OF RESISTANCE

Plants defend themselves against nematodes in several ways.

1. In general, barriers to attraction, the first line of defense, seem to have little importance against nematodes. Infective stages usually enter resistant roots as freely as susceptible ones. *Meloidogyne* juveniles readily enter roots of a resistant alfalfa, but they soon depart to seek better accommodations. A few plants are less attractive to incompatible than to compatible nematodes (3,9).

2. Cells penetrated by nematodes die quickly, thus blocking parasite development. Many plants respond to fungi and bacteria in this way. The term **hypersensitivity** is commonly applied to this mode of resistance.

3. Rates of growth of nematodes in incompatible host – parasite combinations are slower than in compatible combinations. In a good host at favorable temperatures, a *Meloidogyne* female may begin egg production in 3 weeks, but the life cycle of the same species in a resistant host under comparable conditions may be extended to 6 weeks.

4. Resistant plants often inhibit growth of sedentary females and the sex ratio may change in favor of males. In addition, egg production per female is often lower in resistant plants. Consequently nematode populations increase more slowly in fields with resistant cultivars than with susceptible ones.

MECHANISMS OF RESISTANCE

What biochemical mechanisms does a plant have to oppose nematodes? And why is a plant resistant to one pathotype of phytonematodes and not to another? To what signals from a nematode does a plant respond? Does the successful parasite lack compounds that trigger defense responses or does it block them? Is inhibition of parasite growth an active process or is inhibition caused by inadequate nutrition?

Plant pathologists seek answers to such questions for diseases caused by bacteria and fungi; work with nematodes is still to be done. A field of healthy corn or wheat at harvest is witness to the grower's skill in bringing the crop through its season of growth in spite of rodents, birds, insects, nematodes, fungi, protozoa, bacteria, mycoplasmas, and viruses. All compete with us for a share of the stored energy in growing plants. Part of our success comes from the use of pesticides, and part from the plant's own defenses. Plants have well developed defenses against potential parasites, matched by well developed parasite stratagems for circumventing these defenses. Most bacteria and fungi that reach plant surfaces do not grow or invade. They are saprophytes that obtain sustenance from lifeless organic matter. And most soil nematodes are also not phytoparasites; they feed on bacteria, fungi, algae, or animals in soil. The relatively rare successful plant parasite requires genetic information closely matched to that of the plant.

Knowledge of plant defenses is rapidly increasing. A recent review of current research suggests that plants employ various systems to repel and confine parasites and to inhibit parasite growth. The following is adapted from the review (14). Vertebrate animals resist disease by producing antibodies to antigens from the invading organism. These antibodies are highly specific to the antigens that induce them, and they act specifically against the organism. Antibodies to the same antigen in different vertebrates resemble each other to a great extent. Plants, however, produce a great many different antibiotic compounds not highly specific to particular inciting agents or to particular invading organisms. The authors hypothesize that because resistance is the rule and susceptibility to pathogens is rare, "all plants contain the genetic potential for resistance mechanisms to fungal, bacterial, and viral" pathogens. These mechanisms are activated quickly in resistant plants after pathogens appear. Products of altered metabolism inhibit or kill the invaders.

Many plants respond to pathogens by synthesizing low- molecular-weight antimicrobial compounds, **phytoalexins.** These are associated with cell necrosis or other metabolic insults and accumulate in the immediate vicinity of the offending organisms. Phytoalexins are short- lived and are degraded by host and pathogen.

In addition to phytoalexins, plants use other antimicrobial defenses, including:

Preformed toxic molecules;
Physical barriers:
> lignin, a set of large, complex polymers of aromatic acids incorporated into cell walls;
> suberin, the lipid material of cork;
> callose, a complex carbohydrate associated with pores in phloem;
Enzyme inhibitors; and
Agglutinating agents.

"Disease resistance in plants depends on the activation of coordinated, multicomponent defense mechanisms." The key question is, how do pathogens belonging to a particular population elicit defense responses from a host, whereas those from another population invade and damage the same host? Fungi or bacteria that succeed in entering and growing in a susceptible host must be undetected during the early part of an association. Successful pathogens in a particular host may actively inhibit the recognition of their presence. Thus in certain cases, a compatible pathogen in a host suppresses the plant's ability to defend against a subsequent inoculation of an incompatible organism. Conversely, inoculation with an incompatible pathogen may protect a host plant from infection with a compatible pathogen. A possibility now being studied is that pathogens cause release of molecules partly digested from host cell walls and that these in turn activate production of phytoalexins. As understanding of plant defenses increases new approaches to control are suggested.

Plant resistance is a complex phenomenon varying in detail from one host – parasite association to another. Nematodes are much larger organisms than the fungi and bacteria studied by plant pathologists. We do not know enough about phytonematode – host associations to develop testable hypotheses of resistance mechanisms to compare with those of bacterial and fungal infections. The following mechanisms of resistance have been investigated in nematode – plant associations.

Preformed Toxic Molecules.   Certain plants growing in infested soil suppress populations of phytonematodes. *Tagetes patula*, an African marigold, contains nematicidal terthienyl and bithienyl. These compounds also occur in other plants of the family Compositae and probably activate oxygen radicals, that in turn inactivate enzymes and damage cell membranes. They are effective against *Pratylenchus penetrans* and three species of *Meloidogyne*. Root diffusates from some plants in the Cruciferae contain isothiocyanates that inhibit hatching of juveniles from eggs of potato cyst nematodes. Leachates from *Asparagus officinalis* ('Mary Washington') suppress populations of *Paratrichodorus*, probably by inhibiting cholinesterase, an enzyme of nerves and sense organs. Gommers (8) concludes his review of this subject as follows:

> Naturally occurring nematicides discovered so far include alkaloids, phenolics, sesquiterpenes, diterpenes, polyacetylenes, thienyl derivatives and other sulphur containing structures. ... It is difficult to relate the presence of the toxic compounds to *in vivo* nematicidal activity because of insufficient knowledge on the nematicidal mode of action. Therefore, the presence of nematicidal compounds in the plant need not be related to resistance.

However, many investigators have found higher amounts of phenolic compounds in tomato and tobacco resistant to *Meloidogyne* than in susceptible ones. Wild cucumbers resistant to these nematodes are more bitter than susceptible cucumbers. These correlations suggest that preformed compounds may indeed contribute to resistance, but to know this with certainty will require much more research.

Physical Barriers.   Many plants wall off invading pathogens by depositing lignin or suberin in cell walls surrounding the offender. This method of defense is not common in plants resistant to nematodes.

Hypersensitivity.   Many nematodes induce rapid death of cells around them as they enter incompatible hosts. A hungry nematode presumably gets little sustenance from a dead cell.

In certain associations of sedentary endoparasites with resistant plants, feeding sites are induced as in susceptible hosts, but they abort in a few days and the parasite fails to complete its life cycle. We may speculate that plants mobilize defenses to interrupt host – parasite interactions. These defenses may consist of synthesis of antibiotics that inhibit nematode growth. Many infections with *Meloidogyne* or *Heterodera* spp. show high ratios of males to females. Males do not require as much nutrition as females. Their preponderance may indicate interference with full development of feeding sites or full utilization of nutrients by the nematodes.

Phytoalexins.   Several kinds of phytoalexins are known as agents of resistance to fungal and bacterial pathogens, but there are only a few examples of their action against nematodes. Three conditions must be present to prove that phytoalexins contribute to defense against nematodes:

1. They must be synthesized soon after nematode entry (within 4 – 5 days).
2. They must accumulate in cells or tissues close to the pathogen.
3. They must be present in sufficient quantity to damage the nematodes (21).

A study of resistance to *Pratylenchus scribneri* demonstrates these conditions. In a few days after nematodes penetrate roots of lima beans (*Phaseolus lunatus*), high concentrations of two compounds, coumestrol and psoralidin, accumulate at sites of necrosis around the nematodes. The pathogens are paralyzed by concentrations well below those present after infection. Uninfected lima beans also have these compounds but at low concentrations. In the resistant host, nematodes stimulate production of antibiotics that contribute to resistance. The same nematode fails to elicit production of phytoalexins in a susceptible host, snap bean (*P. vulgaris*).

There are two other examples of phytoalexin accumulation in resistant but not in susceptible host – parasite associations (21). A soybean cultivar resistant to *Meloidogyne incognita* accumulates glyceollin in root stelar tissues at concentrations well above the level that immobilizes the nematodes. This phytoalexin does not reach high levels in combinations of a compatible *Meloidogyne* with susceptible cultivars of soybeans. The second example is that of a series of cotton cultivars inoculated with *Meloidogyne* spp. These cotton-nematode combinations range from high to low resistance. Amounts of antibiotics (gossypol and related compounds) localize close to the nematodes and their concentrations range from high to low corresponding to degrees of resistance.

These three examples demonstrate that phytoalexins play some role in plant resistance to nematodes. No doubt other examples will be found. Recent research on nematode surfaces offers intriguing clues to the mechanism of specificity by which some populations of parasites in a host evoke resistance responses; other populations of the same parasite species do not. Restricted sites on nematode surfaces contain molecules that bind to other molecules present in plants. These binding sites on nematodes may be part of the recognition system marking some parasites as foreigners whereas others may escape notice (19).

GENETICS OF RESISTANCE

Genetic control of nematode – plant interactions probably operates in both participants at many stages. Nematodes must be attracted to host tissues, feed at particular places, and derive adequate nutrition for growth and reproduction. Relatively simple associations, such as that between *Trichodorus* and corn, in which little modification of the plant is required for the parasite's success, are likely to involve fewer genes than host – parasite associations of sedentary endoparasites. Cyst nematodes require their hosts to respond in a particular way to their demand for sufficient nutrient throughout their period of growth and reproduction, and the parasite must activate host response throughout the association.

Any disturbance of the course of events in susceptible plants harboring these parasites will interrupt the association. Blocks to parasite development may occur at any point from the initial response to hatching stimulus up to final completion of the female's egg production (1). Both host and parasite must participate actively in this association. When either member ceases its action, or deviates from the path of susceptibility, the whole system deteriorates. We surmise that genetic control operates throughout.

We must distinguish between resistance of a particular cultivar and resistance of a nonhost. A nonhost lacks the combination of characters that a parasite finds necessary. A resistant cultivar or breeding line differs in one or a few qualities from a susceptible host plant of the same species. To develop resistant cultivars, breeders move genes for resistance from a selected plant or from a related species into the genome of desirable crop plants.

Heritable resistance to nematodes has been incorporated into many important crops, including cereals, forages, vegetables, fruits, ornamentals, tobacco, cotton, and soybeans (2,3). Widely planted resistant cultivars exert strong selection pressure on nematodes. In most fields with resistant cultivars there will be a few nematodes that complete their life cycles on the crop. These nematodes probably have the necessary array of genes that permit them to obtain adequate nutrition and to reproduce. Their offspring are more likely to be adapted to the "resistant" host than the original population in that field. In time, most nematodes will be adapted to the formerly resistant cultivar, that will no longer be useful where nematodes are a problem. Thus the work of the breeder is never done.

Almost all determinations of the mode of inheritance of resistance consist of statistical analyses of results of crosses between resistant and susceptible plants. Progeny are tested for resistance at each generation, and the most likely array of genes to account for the ratios is hypothesized. In this way, many monogenic, dominant modes of inheritance have been proposed. Some examples of recessive genes are also known, and some more complex, multigenic types of resistance have been suggested.

Resistance to potato cyst nematodes offers the only example in which inheritance of both nematode and host plant have been studied. *Solanum tuberosum*, the potato, was domesticated in the Andes of Latin America, and cyst nematodes have probably coexisted with this plant since ancient times. Although potatoes were introduced into Europe during the late sixteenth century, they were not used extensively for food until the nineteenth century. Late blight, a disease caused by the fungus *Phytophthora infestans*, destroyed the crop in Ireland in the 1840s, causing great hunger and emigration from Ireland. This catastrophe led to serious study of plant pathology. The first record of potato cyst nematode was published in Germany in 1881. This nematode soon reached notice as a pest throughout Europe. During the two world wars, when potatoes were intensively cultivated, cyst nematodes multiplied to damaging populations in many localities. Potatoes were grown on the same land every year for 20 years in some areas of Great Britain until cyst nematodes reduced yields to low levels during the 1940s (13).

Breeding for resistance began with a survey of the Commonwealth Potato Collection, an assemblage of tuber-forming *Solanum* spp. from the

Andes of South America. Resistance to cyst nematodes surfaced in 1954 in several lines of cultivated *Solanum tuberosum* ssp. *andigena*, in wild *S. vernei*, and in several other wild species (7). Differences among nematode populations were noted soon after introduction of commercial resistant cultivars containing a gene from one of the resistant selections.

Potato cyst nematodes were originally classified as part of the species *Heterodera schachtii*, the beet cyst nematode. In 1923 they were reclassified as a separate species, *H. rostochiensis*, and in 1973 two separate species were established that differed in certain morphological characters and in host range. Currently potato cyst nematodes are classified as *Globodera rostochiensis* and *G. pallida*. These species resemble each other in most respects, but *G. rostochiensis* turns yellow as the female matures whereas *G. pallida* becomes cream colored or white at the same stage. Further, fertile offspring are produced freely in crosses of populations within each species, but the two species do not produce fertile offspring from crosses.

Once sources of resistance had been located, plant breeders began to incorporate them into commercially useful potatoes. Understanding the mode of inheritance was important for this effort. *Solanum tuberosum*, including *S. tuberosum andigena*, is tetraploid, with 48 chromosomes in four sets of 12. A plant bearing two dominant genes for resistance produces gametes in the following proportions:

> one with two dominants
> four with one dominant
> one with no dominants

Mating of two plants of this type should yield F_1 progeny in the following ratios:

> one with four dominants
> eight with three dominants
> twenty-two with two dominants
> four with one dominant
> one with no dominants

This amounts to 35 resistant to 1 susceptible offspring. A cross between a plant with two dominants and another with one dominant gene gives a ratio of 11 resistant to 1 susceptible. And a mating between parents each bearing one dominant gene for resistance should yield offspring in the ratio of three resistant to one susceptible. Plant breeders in the Netherlands crossed a number of potato seedlings grown from resistant selections of

Table 9.1   Segregation in Resistant (Res.) and Susceptible (Susc.) Plants in Progenies Derived from Crosses between Resistant *S. andigenum* Seedlings

| Parentage | Number of Seedlings | | Proposed Ratio | Theoretical | | Proposed Geno types (Dominants only) |
	Res.	Susc.		Res.	Susc.	
CPC[a] 1673 sister lines 2 crosses	70	4	35:1	72	2	HH × HH
CPC 1673 sister lines 8 crosses	291	25	11:1	289	26	HH × H
CPC 1673 sister lines 4 crosses	119	51	3:1	127.5	42.5	H × H

[a] CPC 1673 = a selection from the Commonwealth Potato Collection.

Solanum tuberosum andigena and found progeny that fell very closely into each of the three ratios listed above. Table 9.1 summarizes their results (20).

Toxopeus and Huijsman summarized their results as follows: "Crosses between resistant *andigena* seedlings and between these seedlings and *tuberosum* varieties revealed the existence of a dominant gene H_1 governing the inheritance of resistance." Certain crosses suggested that there are two additional genes, K and L, both of which must be present for resistance.

Subsequent breeding of hybrid potatoes has utilized wild species of *Solanum* as sources of resistance: *S. multidissectum* with a different dominant gene from that of *S. andigena*; crosses of *S. andigena* × *S.multidissectum*, containing both dominant genes; *S. vernei*, believed to have polygenes for resistance; *S. oplocense*; and *S. spegazzinii* (12).

RACES OF PHYTONEMATODES

Although nematologists use the term **race** to designate a population that performs in a particular way on a particular set of differential hosts, this is not the same definition used by others. Zoologists consider that a race is a subspecific group consisting of a population that differs from other groups in the same species by the abundance of various genes making up its collective genotype. Furthermore, the exchange of genes between different races is slower than the rate of exchange within a given race. Thus as a race becomes adapted to its own ecological situation, and as it experiences mutations and other genetic changes, its genetic constitution moves away from that of other races. In other words, subspecific populations diverge as they undergo partial isolation and perhaps different selections. The term **pathotype** refers only to the performance of a population on one or more hosts, and does not express the zoological concept of race.

Plant breeders of cultivars resistant to phytonematodes must take account of genetic variability of the parasites. When resistant cultivars with the H_1 gene were planted in many infested areas, some nematode populations reproduced on them. As more sources of resistance came into general use, a number of different potato cyst nematode populations were recognized in England and Europe, each with its own distinctive reproductive capacity on potatoes resistant to other populations. Table 9.2 lists a set of differential hosts by which eight cyst nematode populations can be differentiated (12). These are labeled Ro1 – Ro5, designating five pathotypes of *G. rostochiensis*, and Pa1 – Pa3, designating three pathotypes of

Table 9.2 Scheme for Recognition of Pathotypes of *Globodera rostochiensis* and *G. pallida*[a]

Clone	Pathotypes[b]							
	Ro1	Ro2	Ro3	Ro4	Ro5	Pa1	Pa2	Pa3
Susceptible *S. tuberosum*	+	+	+	+	+	+	+	+
S. andigena CPC 1673 (gene H_1)	−	+	+	−	+	+	+	+
S. kurtzianum hybrid-KTT 60–21–19	−	−	+	+	+	+	+	+
S. vernei hybrid G-LKS 58.1642/4	−	−	−	+	+	+	+	+
S. vernei hybrid ($VT^n)^2$ 62.33/3	−	−	−	−	±	−	−	+
S. vernei hybrid 65.346/19	−	−	−	−	−	+	+	+
S. multidissectum hybrid P55/7 (gene H_2)	+	+	+	+	+	−	+	+
S. vernei hybrid 69.1377/94	−	−	−	−	−	−	−	−

[a] Reproduced from Ref. (12).
[b] + = abundant cysts; − = few or no cysts; ± = low number of cysts.

G. pallida. Note that susceptible *S. tuberosum* is host to all eight patho-types; that is, it lacks genes for resistance. *S. vernei* hybrid 69.1377/94 is resistant to all populations; that is, it has enough different genes for re-sistance to oppose all populations of potato cyst nematodes found at the time they were tested. The remaining differentials support reproduction of some, but not of all pathotypes.

Comparable variability has been amply demonstrated in other species, including *D. dipsaci, Meloidogyne incognita, M. arenaria*, and *Tylenchulus semipenetrans*. Classification of subspecific variants continues to present difficulties. The simplest method is to establish a set of differential hosts and to test populations against them. The more differentials one uses, the greater the number of subspecific variants revealed (17).

Every bisexual species consists of individuals whose genotypes represent samples drawn from a common pool of genes. At any given locus, more than one allele is likely to be present in the population. This applies to plant hosts as well as to phytonematodes. Therefore, the combined gen-otypes that make up the host – parasite association will vary, and it is not surprising that more than one kind of host specificity is demonstrated among populations of a phytonematode feeding on agricultural crops. Par-thenogenetic species of nematodes also vary, perhaps not to the same degree as bisexual species.

Meloidogyne spp. provide an example of race classification made on hundreds of populations from many origins over the world (11). Six races in the four species of *Meloidogyne* have been distinguished by the North Carolina set of differentials (Table 9.3).

When more differentials are used, more races can be distinguished. Thus four races were separated in *H. glycines* when plant breeding for resistance was developing. But in time new populations that did not fit the preceding classification were found. One experimenter, employing a set of 18 differ-entials, distinguished 36 distinct groups of cyst nematodes among 38 pop-ulations (17). The remarkable feature of the data on *Meloidogyne* is that the four races of *M. incognita* are found in this species from diverse geo-graphic origins.

Genes that confer ability to reproduce on a susceptible cultivar but not on a particular resistant cultivar of a host may be present in high frequency in a population. Resistance is useful against nematodes of that population. But as the population comes under selection pressure of the resistant cul-tivar, frequencies of the relevant genes decline and other alleles become more common. We say that another pathotype is now present, and a new resistant cultivar is needed. Nematodes respond to selection pressures in a few generations. In one study only a few nematodes completed their life cycles after inoculation of resistant tomato with *M. incognita*. Offspring of

**Table 9.3   Usual Response of the Four Common *Meloidogyne*
Species and Their Races to the North Carolina Differential
Host Test[a]**

Meloidogyne Species and Physiological Races	Differential Host Plants[b]					
	A	B	C	D	E	F
M. incognita						
Race 1	−	−	+	+	−	+
Race 2	−	+	+	+	−	+
Race 3	+	−	+	+	−	+
Race 4	+	+	+	+	−	+
M. arenaria						
Race 1	−	+	+	+	+	+
Race 2	−	+	−	+	−	+
M. javanica	+	+	−	+	−	+
M. hapla	−	+	+	−	+	+

[a] Reproduced from Ref. (11).
[b] A: Cotton, Deltapine 61; B:Tobacco, NC 95; C: Pepper, Early California Wonder; D: Watermelon, Charleston Gray; E: Peanut, Florunner; F: Tomato, Rutgers; (−) indicates a resistant host, (+) a susceptible host.

these few were inoculated on the same kind of tomato, and by the third generation a thriving population was reproducing on the tomato breeding line that previously had been a poor host of this species (16).

We need to learn how to measure frequencies of genes for virulence in nematode populations. Some biochemical methods for distinguishing populations of nematodes are under test, including gel electrophoresis to separate proteins and immunological methods to recognize them.

Genes for resistance to a particular nematode do not usually confer resistance against other species of nematodes or fungal and bacterial pathogens. One exception to this statement is that soybeans resistant to *H. glycines* are also resistant to *Rotylenchulus reniformis* but not to other phytonematodes. Moreover, a number of nematode-induced pathologies reduce the genetic ability of plants to resist certain fungal infections. Tobacco plants resistant to *Fusarium* fungi generally grow well in the presence of these pathogens. But when infected with *Meloidogyne* spp. they are vulnerable to infection with *Fusarium* spp. Control of the nematodes often restores genetic resistance of tobacco cultivars to *Fusarium* wilt.

Investigators of the interaction of *Fusarium* wilt fungus and root-knot nematodes in tomato crossed two cultivars, one resistant to both nematode and fungus and the other susceptible to both. One set of F_2 plants was inoculated with nematodes followed by fungus, and another with fungus followed by nematodes. They also took cuttings from each F_2 plant and tested these with nematodes or fungi separately. The F_2 segregated into four groups: 9/16 resistant to both; 3/16 susceptible to the nematode and resistant to the fungus; 3/16 resistant to the nematode and susceptible to the fungus; and 1/16 susceptible to both. This indicates two dominant independent genes, each responsible for resistance to one pathogen. When both pathogens were inoculated on the same plant the ratio became nine resistant to both, three resistant to the nematode and susceptible to the fungus, and four susceptible to both. Resistance to *Fusarium* was lost when root-knot nematodes successfully established an infection. The order in which both pathogens were inoculated did not affect the results (18). These findings differ from those with tobacco in which prior inoculation by nematodes was more effective in breaking resistance to *Fusarium* spp. than simultaneous or subsequent inoculation with nematodes.

CONCLUSION

Parasites have adapted to life within or upon hosts by escaping recognition as intruders and by overcoming host defenses. Correspondingly, hosts have adapted to parasites by reaching an equilibrium that permits some parasites to reproduce but resists destruction of host species by overwhelming parasite attack. As each member of this association evolves in response to the other, there is a continual change in the genetics of the association. Modern agriculture manipulates the genetics of crops, and in so doing runs the risk of severe damage from nematodes, insects, and other organisms. Breeders attempt to replace the slow coevolution of host and parasite by finding and manipulating genes for resistance. As modern agriculture reaches ever greater portions of the earth's arable lands, plant breeders are needed to counter increasing damage from pests. Modern genetic engineering offers promise for more efficient deployment of genes for resistance.

REFERENCES

1. Acedo, J. R., Dropkin, V. H. and Luedders, V. D. 1984. Nematode population attrition and histopathology of *Heterodera glycines*-soybean associations. J. Nematol. 16:48 – 57.
2. Bingefors, S. 1971. Resistance to nematodes and the possible value of induced mutations. In: Mutation Breeding for disease resistance. Int. At. Energy Agency, Vienna, Austria, pp. 209–235.

3. Bingefors, S. 1982. Nature of inherited nematode resistance in plants. In: Pathogens, Vectors, and Plant Diseases: Approaches to Control, (K. F. Harris, and K. Maramorosch, eds.). Academic Press, New York, Chapter 9, pp. 188 – 219.

4. Bingefors, S. and Bengt Eriksson, K. 1968. Some problems connected with resistance breeding against stem nematodes in Sweden. Z. Pflanzenzuecht 59:359 – 375.

5. Boukema, I. W., Reuling, G. T. M.,and Hofman, K. 1984. The reliability of a seedling test for resistance to root-knot nematodes in cucurbits. Rep. Cucumber Genet. Coop., USA 7:92 – 93.

6. Dropkin, V. H. 1959. Varietal responses of soybeans to *Meloidogyne* —a bioassay system for separating races of root-knot nematodes. Phytopathology 49:18 – 23.

7. Ellenby, C. 1954. Tuber forming species and varieties of the genus *Solanum* tested for resistance to the potato root eelworm *Heterodera rostochiensis* Wollenweber. Euphytica 3:195 – 202.

8. Gommers, F. J. 1981. Biochemical interactions between nematodes and plants and their relevance to control. In: Helminthol. Abstr. Series B: Plant Nematol. 50 (1):9 – 24.

9. Griffin, G. D. 1980. Interrelationships of *Meloidogyne hapla* and *Ditylenchus dipsaci* on resistant and susceptible alfalfa. J. Nematol. 12:287 – 293.

10. Halbrendt, J. M. and Dropkin, V. H. 1986. *Heterodera glycines* – Soybean association: a rapid assay using pruned seedlings. J. Nematol. 18:370 – 374.

11. Hartman, K. M. and Sasser, J. N. 1985. Identification of *Meloidogyne* species on the basis of differential host test and perineal- pattern morphology. In: An Advanced Treatise on *Meloidogyne* Vol 2: Methodology. (K. R. Barker, C. C. Carter, and J. N. Sasser, eds.). Dept. of Plant Pathol. North Carolina State Univ., Raleigh, Chapter 5, pp. 69 – 77.

12. Howard, H. W. and Cotten, J. 1978. Nematode-resistant crop plants. In Plant Nematology (J. Southey, ed.). H M Stationery Office, London, Chapter 16, pp. 313 – 326.

13. Jones, F. G. W. 1970. The Control of the potato cyst-nematode. J. R. Soc. Arts 118: 179 – 186.

14. Kuc, J. and Rush, J. S. 1985. Phytoalexins. Arch. Biochem. Biophys. 236:455 – 472.

15. Lauritis, J. A., Rebois, R. V. and Graney, L.S. 1982. Screening soybean for resistance to *Heterodera glycines* Ichinohe using monoxenic cultures. J. Nematol. 14:593 – 594.

16. Riggs, R. D. and Winstead, N. N. 1959. Studies on resistance in tomato to root-knot nematodes and on the occurrence of pathogenic biotypes. Phytopathology 49:716 – 724.

17. Riggs, R. D., Hamblen, M. L., and Rakes, L. 1981. Infra-species variation in reactions to hosts in *Heterodera glycines*. J. Nematol. 13:171 - 179.

18. Sidhu, G. and Webster, J. 1974. Genetics of resistance in the tomato to root-knot nematode-wilt-fungus complex. J. Heredity 65:153 – 156.

19. Spiegel, Y., Cohn, E. and Spiegel, S. 1982. Characterization of Sialyl and Galactosyl residues on the body wall of different plant parasitic nematodes. J. Nematol. 14:33 – 39.

20. Toxopeus, H. J. and Huijsman, C. A. 1953. Breeding for resistance to potato root eelworm. Euphytica 2:180 – 186.

21. Veech, J. A. 1982. Phytoalexins and their role in the resistance of plants to nematodes. J. Nematol. 14:2 – 9.

Ten

CONTROL

INTRODUCTION

Modern phytonematology dates from the discovery of practical compounds to reduce or temporarily eliminate nematodes in agricultural soils. During the 1950s the volatile nematicides (D-D and EDB) produced spectacular improvements in plant growth and crop yields. Control of nematodes is now part of farming in many parts of the world.

Each season crops provide abundant, closely spaced roots, well supplied with water and nutrients—the perfect environment for rapid nematode reproduction. Nematodes are carried by soil, water, wind, and plants. It is virtually impossible to prevent dispersal out of infested fields. After cultivation of sugarcane in the West Indies declined during the nineteenth century, sugar production from beets became a major agricultural activity in Europe. Factories were established in Germany close to beet fields. Beets were lifted from the soils, hauled to factory sites, and dumped in piles until they could be processed. After the beets were used, the remaining soil, contaminated with cysts of *Heterodera schachtii*, was returned to the fields of origin. Although this practice prevented rapid depletion of topsoil, it also recycled inoculum. In many areas sugarbeet production became impossible until control measures were developed.

Reclaimed lands of the Netherlands offer dramatic evidence that phytonematodes move rapidly into unoccupied territory (13). The Dutch have been building dikes for hundreds of years to reclaim land from the sea. Once the wall is secure, seawater is pumped out and drainage tiles are installed. Reeds are planted, rain leaches excess salt from the soil, and then crops are planted. Nematode infestations appear within a few years after this. Damaging concentrations of phytonematodes were found in one area 7 years after reclamation!

How did they get in? Settlers brought trees, shrubs, and perennials from "old" sites for landscaping around homes on the new land. Contaminated soil fell from trucks along the roads. *Meloidogyne naasi*, a species dangerous to cereals, was already found on reeds during early stages of reclamation. Asexual species were the first to cause agricultural problems. Possibly these nematodes can increase rapidly from small numbers whereas sexual species must be present in sufficient numbers to permit frequent matings. Crop sequences and soil types, as well as the host ranges of phytonematode species, were important in the establishment of damaging populations. *Trichodorus teres*, a species with a very wide host range, became generally distributed. However, *Heterodera* spp. depend on specific crops in the rotations and were dispersed more slowly. Rapid invasion of new lands in the Netherlands shows that phytonematodes are easily distributed with soil and plants.

What options does a grower have to keep nematode populations below damaging levels? One can try to avoid bringing nematodes to uninfested lands; starve them out by cultivating poor hosts; use cultural practices such as moisture control or soil amendments to make the environment unfavorable to nematodes; encourage enemies; or use nematicidal compounds. Cost of each antinematode measure must be weighed against probable benefits of market value of increases in yield. As nematode populations rise above thresholds of economic loss, control becomes more difficult and more necessary.

A grower with adequate financial and technical resources can protect crops from losses to nematodes. But farmers without cash or credit have limited means to combat pests. They need genetic resistance for their crops, especially on subsistence farms in tropical agriculture. Nematodes are often devastating where soil temperatures are high, life cycles are short, and costs prevent use of nematicides.

In this chapter we consider the following control options: sanitation, crop rotations, cultural practice, nematicides, and biological control.

SANITATION

Authorities have established quarantines to prevent movement of cysts out of potato fields. They decreed that machinery be cleaned with jets of steam before transport from fields infested with potato cyst nematodes. In practice this has not prevented distribution of cysts because effective quarantines require massive policing (8). One notable success of quarantine is the elimination of *D. destructor* from potatoes in Wisconsin. This nematode was first found in potato fields in 1953. Shortly afterward a strict program of inspection, quarantine, and fumigation with ethylene dibromide was followed. A total of 283 ha of infested and 80 ha of possibly contaminated land was treated. In addition, a state quarantine restricted movement, storage, and sale of tubers from infested fields. Disposal of tubers was closely supervised. Repeated inspections of potatoes grown on treated fields have not revealed any infestation (3).

Sanitation is important in greenhouse and nursery operations. Production of begonia cuttings from mother plants infected with *Aphelenchoides ritzemabosi* distributed the nematodes throughout large greenhouse plantings. Once the problem was recognized, control was achieved. The state of California maintains a sophisticated inspection and diagnostic service to certify that plants sold from nurseries in the state are free of phytonematodes. Imported plants must also pass inspection by quarantine officers posted at the borders and some state lines to prevent introduction of nematodes and other pests.

Plants may be disinfested before shipment or planting. Standard practice in the production of narcissus and other bulbs of flowering plants is to soak them in hot water containing nematicides and fungicides. Time and temperature must be controlled precisely to ensure that *Ditylenchus dipsaci* nematodes are killed without damage to the plants (20). "Both Canadian and U.S. workers recommend pre-soaking bulbs in water, usually at 70 – 80°F (21.1 – 26.6 °C) for two hours just before hot-water treatment." Treatment at 110°F (43.3 °C) for four hours in water containing commercial formalin (1:200) is standard practice. The pre-soak activates dauerlarvae that are otherwise resistant to the treatment. Hot water is also used to kill nematodes on dormant strawberry and chrysanthemum plants.

Nematodes are carried in irrigation water that has drained from an infested field. A measure to prevent dissemination of phytonematodes by irrigation is to hold contaminated water in ponds until the nematodes have settled to the bottom.

CROP ROTATION

Crop rotation is simple in theory. Each species of phytonematode has a range of hosts that may be wide but does not include all crop plants. Nematode populations increase on favorable and decline on unfavorable hosts. A grower need only follow favorable hosts with one or more unfavorable ones to keep nematode populations below damaging levels.

However, the establishment of a successful crop rotation is difficult. Every field of corn or other crop represents a sizable investment of time, energy, and cash. The grower, who usually has machinery and experience for one or a few kinds of plants, favors the same crop year after year. To choose a suitable rotation the grower calculates costs of planting, returns from the harvest, and probable benefit to the main crop. For example, African marigolds contain nematicidal compounds that kill some species of phytonematodes. But there is little or no market for African marigolds. In addition, a candidate rotation cultivar must be adapted to local environmental conditions (temperature, light, soil, wind). And there is always a risk that rotations will enhance multiplication of nematodes that were not hazards to the main crop. Therefore crop rotations must be well tested, usually by universities or research institutes, before they are accepted in practice.

Fallow rotations (that is, maintaining fields free of plants) should also reduce numbers of nematodes. *Pratylenchus thornei* reaches high populations on continuous wheat in dry farming in Israel. Populations of these nematodes are reduced by 90% and the yield of wheat rises by 40 – 70% in experimental rotations alternating wheat with fallow. However, fallow is not always practical because soil may lose productivity and wind and water erosion may be damaging. Moreover, use of unsuitable hosts sometimes causes nematode populations to decline more quickly than fallow rotations.

Trap crops have also been advocated to deplete nematode populations by attracting infective stages. Before life cycles can be completed, the field is plowed and the nematodes die. Trap crops are not widely used because if a grower fails to destroy the crop, nematodes may reproduce before the main crop is planted.

Crop rotations are usually part of an integrated system of control. This consists of rotation together with use of nematicides, resistant cultivars, moisture control, perhaps soil amendment, and in some cases, biological control. The total system is usually called IPM—integrated pest management—and is aimed at all pests, including nematodes.

Population changes under various rotations are illustrated in Table 10.1. This shows 8-year averages of soil samples from three replicated plots on land in North Carolina infested with *Tylenchorhynchus claytoni*. Nematode populations were sampled before planting and after harvest.

The table shows that tobacco supported reproduction of *T. claytoni* to about the same levels after one season regardless of the previous crop. Postharvest populations after tobacco reached levels between 4380 and 7410 nematodes/500 cm³ soil. Over winter these populations declined three- to fivefold to values between 1390 and 1920. Both tobacco and corn were excellent hosts, fescue sod was intermediate, and cotton and peanuts were poor hosts for this nematode. Neither peanut nor cotton maintained populations remaining from the previous tobacco crop. Rates of increase were greatest on a good host when the preplant population was low. In other words, the population on good hosts grew almost to its maximum in one season.

Postharvest populations of *Meloidogyne hapla* in the United States are much higher after peanuts than after tobacco. Fig. 10.1 presents data from

Table 10.1   Seasonal Population Changes of *Tylenchorhynchus claytoni* in Certain Crop Rotation Plots[a]

Crop Sequence	Nematodes/500 cm³ of Soil	
	Preplant	Post-Harvest
Tobacco after tobacco	1920	7410
Corn after corn	1910	5640
Tobacco after corn	2110	7150
Corn after tobacco	1390	7380
Cotton after cotton	270	750
Tobacco after cotton	490	4380
Cotton after tobacco	1740	1540
Peanut after peanut	250	900
Tobacco after peanut	170	5020
Peanut after tobacco	1850	630
Tobacco after fescue sod	1620	7050
Fescue sod after tobacco	1920	3630

[a]Data from Barker and Nusbaum (2).

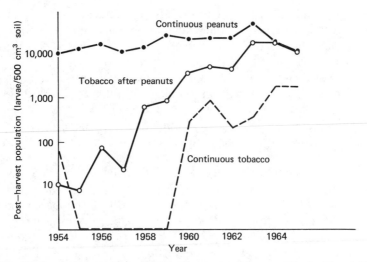

Fig. 10.1 Postharvest populations of *Meloidogyne hapla* larvae at Rocky Mount, North Carolina. Mean values from three replicated plots for continuous peanuts and continuous tobacco and from nine rotated tobacco plots each year. [Plotted from data of Ref. (2).]

a field plot experiment to monitor population fluctuations under monoculture cropping. The field had been in continuous peanuts for 16 years prior to 1954. There were three replicated plots of continuous peanuts, three of continuous tobacco, and nine of tobacco rotated yearly with peanuts. Continuous peanuts maintained populations fluctuating between 8675 and 38,517 juveniles/500 cm³ of soil after harvest. For 6 years the population of *M. hapla* hardly reproduced on continuous tobacco, but from 1960 onward, this population slowly increased on tobacco. It never attained the high levels found after peanut. Perhaps monoculture selected a pathotype adapted to tobacco that was previously present in very low proportion in the soil. The graph of *M. hapla* on tobacco rotated with peanut shows the same change. From 1954 to 1957, the nematodes barely survived on tobacco, but from 1958 onward, populations under tobacco rose steadily to values close to those under peanuts.

Threshold densities for damage by a species of nematode vary from one host to the next. *T. claytoni* at high densities does not damage tobacco but populations greater than 100/500 cm³ of soil before planting cause decline of azaleas. A field with *Pratylenchus penetrans* is dangerous for potatoes at 500 or more/500 g of soil but 1 – 5 *P. penetrans*/500 g are sufficient to harm daffodils. Threshold densities for damage also vary from one ne-

Table 10.2   Populations of Three Phytonematodes after Five Years of Monoculture of Various Plants used to Depress *M. incognita* (Nematodes/500 cm³Soil)[a]

Plants	*Trichodorus christiei*	*Xiphinema americanum*	*Pratylenchus brachyurus*
Fallow	333 (weeds?)	0	7
Bahia grass	1632	326	13
Crotalaria	163	2504	386
Bermuda grass	9254	353	203
Marigold	6540	80	3

[a]Source: Murphy et al (16).

matode to another on the same host. Tobacco is harmed by 25 – 50 *M. incognita*/500 g, by more than 200 *P. brachyurus*/500 g, and not by any levels of *T. claytoni*.

Rotation crops must not only protect the main crop and at the same time be marketable, but must also avoid increase of troublesome nematodes. Table 10.2 lists populations of ectoparasites after 5 years of monoculture of several nonhosts of *M. incognita* used in rotations in southeast United States. Some species of ectoparasites attain high populations on these plants.

CULTURAL PRACTICES

A grower can sometimes make simple adjustments to avoid severe damage from nematodes.

Early Planting.   Potatoes and beets begin growth in cool soils; early plantings permit both crops to get started before nematodes are active. When the soil warms and juveniles of cyst nematodes invade growing roots, plants have already accumulated reserves and can withstand attack. Nematodes, like other forms of life, are not easily thwarted! Continued early plantings may select ecotypes of phytonematodes adapted to low temperatures.

Drying.   Nematodes may be controlled in hot, dry climates by plowing the soil before planting, thus exposing them to heat and drying. Plowing fields several times about 2 weeks before sowing reduces nematode populations in India. The soil must be turned to bring nematodes from deeper layers to the surface.

Moisture Control. Agriculture in desert lands of California depends on precise applications of carefully measured amounts of water. Need for irrigation is determined by instruments installed in the field to measure moisture tensions. A year-long study of a citrus orchard on desert soil showed a cycle of abundance of *Hemicycliophora arenaria* (21). Nematode numbers were highest in January (2900/500 cm^3 of soil), declined to a low (<100/500 cm^3) in August, and rose again in December. The authors concluded from greenhouse experiments that soil aeration was the critical factor regulating nematode reproduction. High soil moisture impedes oxygen diffusion and depresses this nematode's ability to build large populations in orchards on desert soils. From November through February growers applied 5 acre-in. of water at 13-day intervals and nematode populations rose. Irrigation increased to about 8 acre-in. at 7-day intervals in March – May, and to 13 –18 acre-in. at 3-day intervals in June – August and populations declined. Then irrigation declined through September – October to the amounts applied in March – May and continued declining until November.

Burning. Producers of grass for seed production in the Pacific Northwest of the United States set fires each year after harvest to remove crop residues and to control *Anguina* and fungal diseases. The same practice is used to improve pastures elsewhere. Although fire destroys aboveground nematodes, it does not heat soil sufficiently below the uppermost layers to affect root parasites (6).

Soil Amendment. Early literature on nematode control stressed the importance of maintaining high organic content in soils. Many substances can be added to soil to increase organic matter: manure from domestic animals, sewage sludge from municipal waste disposal facilities, crop residues after harvest, and cover crops. Processing of agricultural products produces organic matter suitable for incorporation into soil: cottonseed hulls, bagasse (pressed cane from sugar mills), and oil cakes (pressed oilseeds). Traditional agriculture in the Orient maintains high organic content of soils with such materials. Nematologists are now examining use of these by-products and other soil amendments for nematode control.

Many organisms, from bacteria to earthworms, incorporate organic matter into soils. This ultimately forms humus. Soils with adequate humus have excellent texture and good water-holding capacity, and are generally more favorable to plants than soils without sufficient organic matter. Therefore, plants grow well in such soils and are not under as much stress as those in poorer soils. Breakdown of organic matter releases compounds that may be toxic to nematodes. In particular, decomposing residues of plant tissues release simple organic acids such as acetic, propionic, and butyric acids. These may remain for several weeks in concentrations sufficient to kill some phytonematodes but are not toxic to the free- living species (19). Bacteria produce hydrogen sulfide from decomposing organic matter under anaerobic conditions in flooded rice fields. This gas kills nematodes.

Addition of organic matter to soil may result in rapid changes. Numbers and kinds of microorganisms reflect abundance and kinds of substrate on which they feed. Chitin derived from crustaceans, for example, stimulates growth of **actinomycetes**, some of which are antagonistic to phytonematodes. It also stimulates increase of fungi with enzymes capable of digesting chitin. Many of these fungi penetrate cyst nematodes to attack chitinous walls of eggs. During degradation of organic matter by microorganisms, toxic metabolites that kill nematodes are released. In addition to toxins from this source, many plants contain nematicidal compounds that are released from amendments such as oil cakes.

Organic matter may also increase abundance of **predatory fungi**. In this case, bacteria multiply on organic matter, nematodes that feed on bacteria multiply, and fungi that feed on nematodes also increase their abundance. Phytonematodes may decline as a result. Some recent work on effect of soil amendments is reviewed here.

Complexity of soil life is illustrated in pot experiments to test effects of chitin amendments (17). Chitinolytic fungi increased with addition of up to 2% chitin (weight of amendment/weight of soil). Numbers of actinomycetes also increased with addition of up to 3% chitin. At higher concentrations, numbers of both organisms did not change, or they declined. Probably toxic metabolites of chitin degradation were present. Bacteria also increased in soils with up to 2% chitin, but not at higher concentrations. Numbers of nematodes parasitizing soybean in amended soil were reduced; however, not all species were equally affected. *Helicotylenchus dihystera*, an ectoparasite, was less sensitive to chitin amendments than *Heterodera glycines*, *Meloidogyne incognita*, or *Tylenchorhynchus claytoni*. And non-phytoparasitic species reached high levels in chitin-amended soil. Chitin inhibited growth of soybean roots, but the experiments terminated before plants reached maturity and no conclusions on yield could be drawn (17). Although these results do not rigorously prove that the microorganisms damaged nematodes, they do show how complex the soil biota is and they suggest that useful results should follow from comprehensive studies of soil amendments.

Cowpeas (*Vigna unguiculata*), an important crop in Africa, are severely limited by *Meloidogyne* spp. Nematicides are effective, but beyond the reach of most producers. Amendments with crop residues for control of rootknot in Nigerian cowpea production (3) were compared with carbofuran, a nematicide, or NPK, a fertilizer. The experiments were made both in a greenhouse and outdoors in replicated plots (2 m × 1 m). "Although based on limited data, projected economic returns show that the use of cocoa pod husks and cassava peelings generated the highest returns per hectare in monetary values when compared with the other treatments." Rice husks, however, carried pathogenic fungi to the cowpeas, and cocoa pod husks probably improved soil tilth as well as contributing nutrients. Addition of cocoa pod husks and cassava peelings as well as rice husks resulted in reductions of *Meloidogyne incognita* populations, but the mechanism was not established (3).

Certain plants grown in India for their oil-rich seeds have nematicidal properties. Among these are neem (*Azadirachta indica*), mahua (*Madhuca indica*), castor oil plant (*Ricinus communis*), and peanut (*Arachis hypogea*). After seeds have been pressed to extract oil, cakes of residues are added to soil. They protect highly susceptible plants from *Meloidogyne* spp. and other nematodes. These oil cakes have phenolic compounds that may be the active agents against nematodes (1).

Solar Pasteurization. Horticulturists have known for many years that maintaining soil at 50 – 60°C (pasteurization) improves its qualities for plant growth. Until recently it was not possible to apply this knowledge to field cultivation. But now that transparent plastic is available at reasonable cost it is practical to take advantage of energy from the sun to pasteurize soil in the field. In Israel, plastic mulch applied over soil before plants were in the ground reduced fungal pests and killed weeds (9,10). Soil temperatures under transparent plastic in the Jordan Valley during the hot summer reached a maximum of 52°C at a depth of five cm and 42°C at 15 cm. The cover was left in place for 4 – 5 weeks and plots were kept moist by drip irrigation to improve heat transfer. Similar experiments in the cooler climate of New York proved that numbers of hatched juvenile *Globodera rostochiensis* were reduced under plastic mulch. Numbers of viable eggs within cysts were also reduced (14). Pest populations diminished in layers of soil where temperatures were well below known lethal levels. Enhanced biological control by microorganisms and weakening of pathogens probably play major roles in beneficial effects of solar heating. Plastic mulches will become part of total control programs to supplement other measures. They are cheap, leave no objectionable residues of poisons in soil, and can serve where chemical control is too expensive to be useful.

NEMATICIDES

Investigations of control methods were undertaken immediately when nematodes became known as agricultural pests in the mid-nineteenth century. Many attempts were made to use chemicals and soil amendments to improve beet cultivation. Carbon disulfide, extracts of decomposing cadavers, and other materials were tried. But no outstanding advances took place until chloropicrin, a war gas, became available as surplus after World War I. This compound, used as a fumigant, improved crop production where poor yields followed repeated monoculture. Its outstanding success stimulated search for other compounds, including those effective against nematodes. With discovery of the nematicidal effect of D-D, a by-product of petroleum refining, and of EDB, ethylene dibromide, the modern development of nematicides was underway.

Nematicides fall into two groups, fumigants and nonfumigants.

Fumigants are volatile liquids that vaporize and dissolve in the soil solution. Their high vapor pressure distributes the gas in all directions through pores. Two types of fumigants are in use:

1. Halogenated aliphatic hydrocarbons such as ethylene dibromide and 1,3-dichloropropene (1–3 D), used at rates active against nematodes and soil arthropods.
2. Liberators of methylisothiocyanates such as metham sodium, dazomet, and methylisothiocyanate itself. Together with methyl bromide and chloropicrin, these are general biocides.

Nonfumigant nematicides are also of two types:

1. Organophosphates such as fenamiphos, thionazin, ethoprop, fensulfothion.
2. Carbamates such as oxamyl, aldicarb, carbofuran, tirpate, metham sodium, and methomyl.

Species of phytonematodes differ in their sensitivity to nematicides. DBCP, a halogenated hydrocarbon now taken off the market because it inhibits human male fertility, is less effective on *Trichodorus* than on other species. EDB is less toxic to *Aphelenchus avenae* than to *Tylenchulus semipenetrans* or *Meloidogyne javanica*. Within a single species, some stages are more sensitive than others. Juveniles die at lower concentrations of fumigants than adults; nematodes undergoing molts are more sensitive than stages between molts. Eggs are less sensitive than juveniles or adults.

And nematodes adapted to dry conditions are less sensitive to nematicides than those under moist conditions. To compare the efficacy of nematicides, the product of concentration × time of exposure is taken. A high concentration for a short time is as effective as a lower concentration for a longer time.

Mechanisms of action of nematicides appear to be of two kinds. Individuals exposed to halogenated hydrocarbons at sublethal doses (concentrations × time) become hyperactive, then cease movement and fail to respond to stimuli. When returned to nematicide-free fluid, they recover. They die at higher concentrations. EDB is believed to interfere with oxidation of Fe^{2+} in the cytochrome chain, leading to a reduction in production of ATP, an energy-rich compound generated during metabolism. It also alkylates amino acid components of proteins, which are then hydrolyzed. This process blocks action of esterase and protease enzymes, ultimately killing the nematodes.

Carbamates and organophosphates inhibit cholinesterase activity, resulting in failure to regulate acetylcholine, a neurotransmitter. This causes paralysis and loss of sensory perception, but is not immediately lethal. Nematodes recover after removal of the pesticides (22). Disturbance of neurotransmission interferes with life cycles of phytonematodes. This blocks feeding in some species, thereby affecting virus transmission; it also hinders normal growth of parasites already inside plants. Various compounds protect roots against invasion. At low dosages, nematicides may disturb sensory function and inhibit nematodes from finding roots, or each other. Organophosphates and carbamates differ in their effects on particular species of nematodes.

Persistence of nematicides in soil after application is important in contamination of the environment. Some compounds are readily destroyed but others are more persistent. Aldicarb, a highly toxic pesticide, converts to aldicarb sulfoxide and aldicarb sulfone in soil and in plants. All three are taken up by roots and translocated to all parts of plants. Studies in vineyards show that all three compounds can be detected throughout grapevines, including the fruits (5). Concentrations gradually decline as the grapes mature and toxic residues in grapes at harvest are less than 1 ppm. Carrots at harvest may contain higher levels of these compounds, probably because carrots mature more quickly than grapes. The half-life of aldicarb in soil and in carrots is estimated to be about 14 days.

The public is understandably concerned about contamination of ground water by toxic compounds. In environments favoring rapid destruction of aldicarb, as in Florida citrus groves, small amounts of this nematicide or its derivatives are completely degraded before reaching groundwater. But under certain conditions, degradation proceeds more slowly and residues can reach potable groundwaters. These conditions include high usage, highly permeable acidic soils, high water recharge rate just after application, low soil microbiological activity, and shallow potable groundwater (7). DBCP was in use throughout California until 1977. It resulted in widespread contamination of groundwater underlying major use areas. Now oxamyl and other compounds have replaced this nematicide. The half-life of oxamyl is 1 – 4 days in desert soils under irrigation and the compound does not accumulate in soil or groundwater even after repeated treatments.

Efficacy depends on proper application of nematicides to soils at favorable times. Each compound has its own characteristic solubility in water, vapor pressure, and ease of degradation by microorganisms, as well as toxicity to nematodes. Effective applications are based on these characteristics as well as on soil conditions. Thus fumigants should be applied deeply enough to avoid escape to the atmosphere and the soil should have enough pores free of water to permit adequate distribution of the toxic compound. Figure 10.2 shows a typical arrangement for application of fumigant. Special equipment to pack the surface after fumigation is often used. Because fumigants escape to the atmosphere, the upper few centimeters of soil may not contain sufficient concentrations of nematicide for a sufficient time to kill nematodes. Additional treatments may be required.

Fig. 10.2 Photo of equipment to dispense fumigant into soil. Note tubing leading to each of the chisels through which fumigant flows from reservoir (not shown). (Photo courtesy of R. Rodriguez-Kabana.)

Soils of high organic content are more difficult to fumigate successfully
because nematicides may be bound to the surface of organic particles.
Some treatments are phytotoxic and thus require time for the soil to aerate
before planting. Compounds of low vapor pressure are usually applied as
granules mixed with soil or are added via irrigation. Figure 10.3 shows the
dramatic improvement of crop growth resulting from treatment with a
nematicide.

Nematologists must have reliable information from replicated field experiments to recommend best use of nematicides for yield improvement.
A comparison of *Meloidogyne arenaria* populations and yields of soybeans
and peanuts after various methods of nematicide application illustrates the
kind of work required (18). The investigators applied nematicides at several
rates over seeds at planting. Eight replicated plots arranged in a randomized
design were established for each treatment in a large field with a high
population of nematodes. In the soybean experiments, four nematicides
were applied as granules at two rates using three methods. Samples for
counts of nematodes were taken from each plot just before harvest and
yield of each plot was recorded. The experiment included 192 plots. Similar
trials on peanuts totaled 288 plots. Table 10.3, reproduced from Ref. 18,
illustrates the statistical analysis used to establish validity of the results.
The authors concluded that banded applications were superior to in-furrow

Fig. 10.3 Effect of nematicide treatment on growth of crop. The row with poor growth
was not treated. (Photo courtesy of H. Rhoades.)

Table 10.3 Effect of Method of Application on Efficacy of Temik 15G (aldicarb) against *Meloidogyne arenaria* in 1980 Field Experiment with Florunner Peanuts at Headland, Alabama[a]

| | Pounds of Active Ingredient per Acre | | | | | |
| | One lb | | Two lb | | Three lb | |
Method of Application	Larvae[b]	Yield (lb/A)	Larvae	Yield (lb/A)	Larvae	Yield (lb/A)
In-furrow	102 A[c]	2335 C	42 B	2664 BC	41 B	3027 B
5-in. band	66 B	2819 B	72 A	3417 A	10 C	3189 B
7-in. band	44 BC	2957 B	33 BD	3352 A	14 C	3221 AB
14-in. band	25 C	2993 B	21 BD	2863 B	30 BC	3497 A
Control	96 A	2449 C	96 A	2449 C	96 A	2449 C
Soilbrom[d]	15 D	3359 A	15 D	3359 A	15 C	3359 A

[a] From Rodriguez Kabana et al (18).
[b] per 100 cm^3 soil.
[c] Figures for variables are the averages of eight replications; those within the same column followed by a common letter were not significantly different (**P** = .05).
[d] Copyright label.

applications both for control of the nematode and for consequent yield increase. Band widths of 12.7 cm (5 in.) or 17.8 cm (7 in.) were adequate; wider bands offered no advantage. They also surmised that

> systemic nematicides . . . are most effective when applied to soil in a manner that enhances their quick absorption by developing root systems. In this connection, diffusion of the nematicide through soil from the place of application into the zone where roots develop is critical.

Growers in technically advanced agriculture routinely apply more than one pesticide to a crop during the growing season. Does use of an herbicide alter efficacy of a nematicide? Data from experiments on this question indicate complex interactions of the chemicals. Some combinations appear to be without influence, others depress nematicidal efficacy, and still others appear to enhance it.

Nematicides are designed to block essential processes in nematode metabolism. Because these same processes operate in other organisms including humans, nematicides are inherently dangerous compounds. Table 10.4 contains the principal nematicides in use in the United States arranged in order of increasing oral toxicity as measured by the amount that will kill half the animals exposed to them. Although toxicity to humans is not

Table 10.4   Acute Oral and Dermal Toxicity of Nematicides (mg of compound/kg of body weight)[a]

	Chemical Name	LD_{50} Oral[b]	LD_{50} Dermal[b]
	Fumigants		
Metham	Sodium methyldithiocarbamate	820	(rb) 800
MBR	Methyl bromide	200[c]	
DD	1,3-Dichloropropene and related chlorinated C_3 hydrocarbons	140	2100
	Organophosphates		
Diazinon	O,O-Diethyl O-[6-methyl–2-(1-methyl-ethyl)-4-pyrimidinyl] phosphorothioate	66 – 600	379 – 1107
Ethoprop	O-Ethyl S,S-dipropylphosphorodithioate	61	26 (rb)
Phenamiphos	Ethyl 4-(methylthio)-**m**-tolyl isopropylphos-phorodithioate	8 – 10	178 – 225 (rb)
Demeton	Mixture of O,O-diethylS(and O)-[2-(ethyl-thio)ethyl]phosphorothioates	2 – 12	8 – 200
Terbufos	S-[(1,1-Dimethylethyl)thiomethyl] O,O-diethylphosphorodithioate	4 – 9	1 (rb)
Phorate	O,O-Diethyl S-[(ethylthio)methyl] phospho-rodithioate	1 – 5	2 – 7
	Carbamates		
Carbofuran	2,3-Dihydro–2,2-dimethyl-7-benzo-furanylmethylcarbamate	5	885 (rb)
Oxamyl	Methyl N′,N′-dimethyl-N-[(methylcarba-moyl)oxy]-1-thiooxamimidate	5	—
Aldicarb	2-Methyl–2-(methylthio)propionaldehyde O-(methylcarbamoyl)oxime	1	5 (rb)

[a] Data from McGrath, H., Feldmesser, J. and Young, L. D., eds. 1986. Guidelines for the Control of Plant Diseases and Nematodes, U.S. Agricultural Research Service, Handbook No. 656. Superintendent of Documents, U.S. Government Printing Office, Washington, DC., 274 pp.
[b] Acute oral (AO) and acute dermal (AD) levels expressed as LD_{50} values for white rats except for rabbit (rb).
[c] Acute vapor toxicity for man in ppm.

measured precisely by exposing white rats or rabbits to these compounds, the data are an appropriate guide to their danger.

Nematicides depress nematode populations and have additional effects. General biocides such as methyl bromide kill soil fauna and liberate organic compounds, including nitrogen, that temporarily increase soil fertility.

Some nematicides (for example, carbofuran) appear to have favorable effects on plant growth and improve nitrogen fixation by nodulating bacteria of legumes.

BIOLOGICAL CONTROL

Yields usually decline during continuous cultivation of the same crop for many years in the same field. Populations of pathogens increase to damaging levels, and in many instances the soil becomes totally unsuitable for the crop. Before plant pathologists described the major pathogens, people said a soil became "tired" of its crop. But this situation is not universally true. In some places, for example, certain fields on the island of Maui where pineapples have been grown year after year, good yields continue.

Phytonematodes themselves are food for other organisms. "Suppressive soils" contain many parasites and predators of pathogens. Continuous cultivation of a susceptible crop permits phytonematodes to multiply, and in turn these parasites stimulate their enemies to increase. An equilibrium may be reached that protects the crop from low yields caused by nematodes. Other pathogens may be subject to suppression in the same way. Once a few cases of suppressive soils were known, the possibility of biological control became apparent. Can we favor the enemies of nematodes? Can we introduce predatory fungi in sufficient numbers to control phytonematodes in the field? And will such introduced organisms persist to protect a crop for more than one year? Can a field be managed in a way to encourage biological control? Answers to such questions do not come easily.

Interest in parasites and predators of nematodes is growing as use of pesticides in intensive agriculture faces increasing opposition from the public. Parasites of nematodes include an intracellular rickettsia-like organism of unknown effect, *Bacillus penetrans* that kills, and various fungi that penetrate and consume nematodes. Predators include carnivorous nematodes, trapping fungi of various kinds, mites, and Collembola. Thus a nematode leads a precarious existence in soil, having to thread its way past predators and parasites en route to a growing root. A species survives by reproducing in sufficient quantity to ensure that some will continue the next generation.

Biological control denotes the reduction of nematode populations by parasites and predators of eggs, juveniles, or adult nematodes. Humans have exerted a kind of biological control by breeding resistant plants and by our cultural practices. But the term "biological control" refers specifically to parasites and predators.

The first experimental evidence of biological control of nematodes appeared when investigators were preparing plots for tests of fertilizers. They drenched soil with solutions of formaldehyde to eliminate root disease. To their surprise, populations of cyst nematodes increased dramatically on cereals planted in treated plots. Formaldehyde had suppressed fungi that were limiting increases of *Heterodera avenae*, the cereal cyst nematode. This finding aroused interest in predators and parasites as possible agents of control.

H. avenae reproduces on grasses including cereals (oats, wheat, barley, and corn). When pastures are converted to cereal production, populations of cyst nematodes increase markedly above thresholds of damage. However, continued cultivation of cereals may reduce nematodes to low levels (11,12). Several species of fungi attack these pests in British soils and elsewhere in the world. Motile spores of one species invade females soon after they break through roots from the interior to the surface, mycelium grows through the entire cyst, and thick-walled, persistent spores are produced. These disperse through soil when infected cysts disintegrate. Other species of fungi prefer eggs and still others trap motile juveniles.

Soil conditions affect activity of soil fungi. Results of field trials show the difference between a year with adequate moisture and one without (Table 10.5).

During dry conditions in 1976, the fungi were inactive but there was enough moisture to enable cysts to increase. But in 1975, moisture was sufficient to permit active fungi to suppress the nematodes. Many fungi, in addition to the ones affecting *H. avenae*, invade nematodes. There is no evidence that they protect crops against populations of phytonematodes.

Fungi specialized to capture and consume nematodes are the product of extensive evolution. Trap – forming species of fungi have attracted many investigators both because of their fascinating adaptations to predation on nematodes and for their potential as biological control agents. One type produces rings of a few cells. When a nematode touches the inside of this trap, the cells swell in a fraction of a second and the nematode is caught.

Table 10.5   Rates of Increase of Cereal Cyst Nematodes Affected by Moisture in Two Successive Years[a]

Year	Oats	Wheat	Barley	Rainfall (mm) (May + June + July)
1975	0.8	0.6	0.2	126
1976	4.1	3.2	1.5	66

[a] Adapted from Kerry (11).

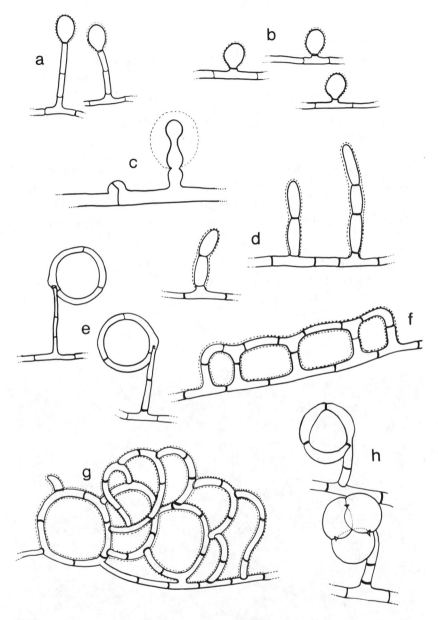

Fig. 10.4 Organs of capture in predatory nematode-destroying fungi. (**a**) stalked adhesive knobs; (**b**) sessile adhesive knobs; (**c**) hour-glass adhesive knobs of *Nematoctonus*; (**d**) adhesive branches; (**e**) nonconstricting rings; (**f**) two-dimensional scalariform adhesive net; (**g**) three-dimensional adhesive net; (**h**) constricting rings. (From G. L. Barron. 1977. The nematode-destroying fungi. Topics in Mycobiology No.1, Canadian Biological Publications, Guelph, 140 pp., with permission.)

Hyphae then penetrate through the body wall and kill the victim. These fungi apparently exist in soil as slow-growing saprophytes with few or no traps. In the presence of nematodes traps are formed. Another kind of trap is a three-dimensional mesh coated with mucilaginous material that binds nematodes. Recent research has focussed on surface characteristics of nematodes and of fungi that join the two together. Figures 10.4 – 10.8 illustrate aspects of the biology of fungal predators and parasites of nematodes.

Bacillus penetrans is a parasite known for many years in association with *Meloidogyne* spp. Spores, sometimes in great numbers, attach to infective juveniles and produce a germ tube that penetrates the body (Fig. 10.9). The organism proliferates into daughter colonies, eventually filling adult females with spores subsequently released to soil. Spores do not actively seek another nematode, but are dispersed by water movement or soil disturbance. This parasite appears to have great potential for use as a biological control agent. Spores persist in soil, are not killed by nematicides. Under continued cultivation of hosts susceptible to rootknot, a soil may accumulate enough spores to decimate the nematodes (15).

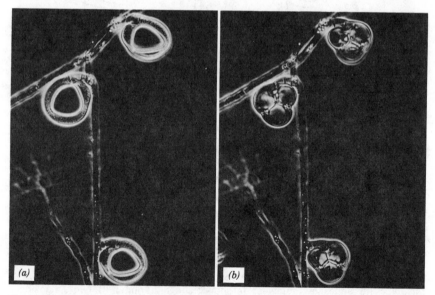

Fig. 10.5 Constricting rings. (a) normal rings; (b) rings triggered by heat to constrict. (From G. L. Barron. 1977. The Nematode-destroying Fungi. Topics in Mycobiology No.1, Canadian Biological Publications, Guelph, 140 pp., with permission.)

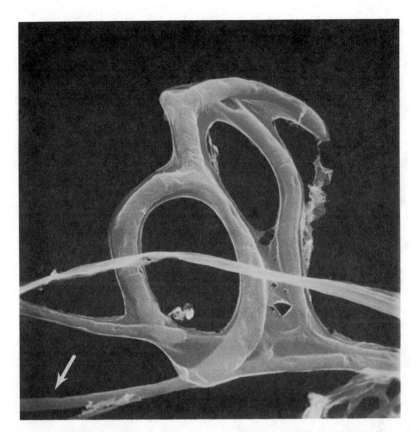

Fig. 10.6 Scanning electron micrograph of part of net of *Arthrobotrys oligospora* showing adhesive coating over surface of net (×2000). Arrow indicates hypha without adhesive coating. (From B. Nordbring-Hertz. 1972. Scanning electron microscopy of the nematode trapping organs in *Arthrobotrys oligospora*. Physiol. Plant. 26:279–284, by permission.)

Mankau, in an excellent review of biological control, points out that although the efficacy of *B. penetrans* has been proved in the field, we are still far from using it as a substitute for nematicides. Biological strains exist—some spores that attach to *Meloidogyne* spp. do not affect *Pratylenchus* spp. and vice versa. The parasite occurs in many localities over the world. Undoubtedly extensive surveys would uncover some especially effective strains specific for particular phytonematodes. Not all spores that attach to a juvenile nematode are lethal. Those that enter are ineffective until the adult stage that succumbs to the parasite is reached. The gap in our knowledge at present is a lack of methods to cultivate *B. penetrans* in

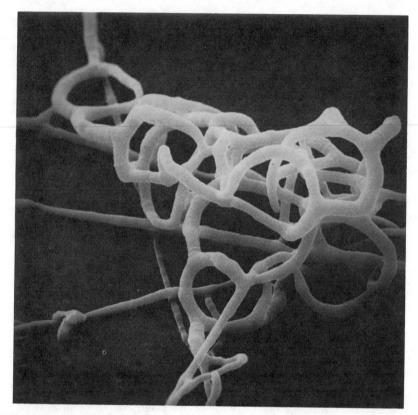

Fig. 10.7 Scanning electron micrograph of net of *Arthrobotrys oligospora* (×1200). (From G. Lysck and B. Nordbring-Hertz. 1983. Die Biologie nematodenfangender Pilze. Forum Mikrobiol. 6:201 – 208, by permission.)

artificial media, thus permitting large scale production and distribution of spores in the field (15). Presently spores for inoculation can be produced by exposing juveniles of *Meloidogyne* to spore suspensions in water. Spore populations can then be increased by inoculating infected juveniles on susceptible hosts. At harvest, roots with spore-bearing nematodes can be dried and ground to fine particles to produce spore suspensions. Such suspensions will protect plants in the field from depredation by *Meloidogyne*. This method is too laborious for commercial purposes.

Efforts to achieve sustained biological control of nematodes in the field have not succeeded. Soil is a powerful buffer against our efforts.

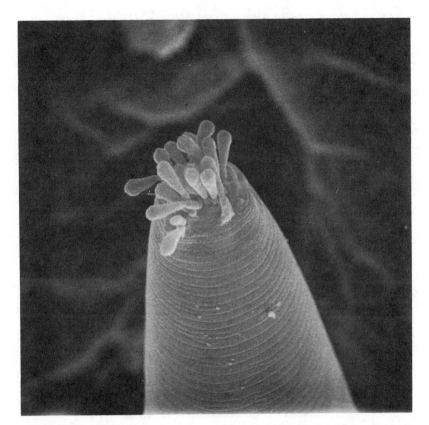

Fig. 10.8 Scanning electron micrograph of the mouth region of the nematode *Panagrellus redivivus* with conidia of the endoparasitic nematophagous fungus *Meria coniospora*. (From H-B. Jannson. 1982. Attraction of nematodes to endoparasitic nematophagous fungi. Trans. Br. Mycol. Soc. 79:25 – 29, by permission.)

An average fertile field soil contains approximately 10^9 bacteria, $10^5 - 10^8$ actinomycetes, $10^5 - 10^6$ fungi, and $10^4 - 10^5$ protozoa reproductive units per gram, plus a large number of fungivorous nematodes, tardigrades, Collembola, mites, and other assorted micro- and meiofauna. It is difficult to believe that adding one additional organism will have an immediate measurable effect (15).

Work in this area until now is largely descriptive and scattered. Sustained investigations of suppressive soils coupled with detailed studies of the ecology of control agents may ultimately produce useful results. Modern tech-

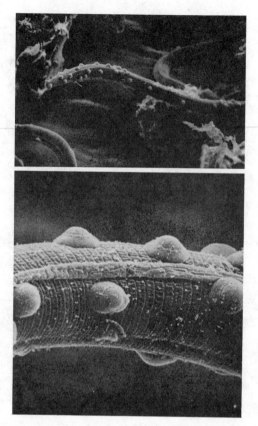

Fig. 10.9 Scanning electron micrograph of a juvenile *Meloidogyne incognita* showing spores of *Bacillus penetrans* attached to the cuticle. The parasite grows through the cuticle and invades the nematode's interior. (Courtesy of W. Wergin and R. Sayre, U.S. Department of Agriculture, Nematology Laboratory.)

niques of biological engineering look especially promising. Metabolic products toxic to nematodes from soil organisms and aggressive behavior toward prey, as well as other desirable characteristics are under genetic control, and we may anticipate that they can be assembled into marketable organisms for control measures to supplement or replace conventional nematicides.

Nematicidal Plants

Investigators in Holland found that flowering bulbs could be protected from damage by *Pratylenchus* by cultivation of African marigolds (*Tagetes*).

These plants contain compounds that stimulate production of oxygen radicals, thus blocking steps in nematode metabolism (4). The possibility of control by use of nematicidal plants has stimulated research on the occurrence of such plants and on the biochemistry of the antinematode action. Several species, in addition to marigolds, are known to have nematicidal properties. These include species of Compositae, cultivated asparagus, *Crotalaria* spp., pangola grass, and the plants mentioned earlier in the discussion of oil cakes used as soil amendments. It has not been practical until now to employ nematicidal plants on a large scale for nematode control.

CONCLUSION

When nematicides first became available, nematologists sought methods of eradicating nematodes. But eradication proved unreachable. Now we strive to depress nematode populations below the level of damage, knowing that reproductive potential of these parasites will bring populations back to high levels. Therefore protection of young plants is important to permit crops to grow before nematodes can reach damaging concentrations. We also try in various ways to oppose population increases. Ultimately we must accept nematodes as part of crop production and strive to keep the tax we pay to them at a minimum. Because agriculture is diverse in different regions, with different crops, different cultural practices, and different climates, we must maintain a high level of communication to share knowledge of nematode control.

Numbers of human beings continue to increase. In common with all forms of life, we push against the limits of our environment. One of the limits is agricultural production. Most of the easily managed land is already under cultivation. More of the unsuitable areas are also being used for farming. We cannot tolerate crop yield losses from failure to control parasites of plants.

Insects, fungi, nematodes, bacteria, and viruses do not respect political boundaries. We must organize efforts on an international scale. First steps in this direction are already being taken: the international *Meloidogyne* project; international quarantines; centers for research on specific crops, such as for potato research in Peru, rice research in the Philippines, and others; international conferences; and sales organizations of nematicide producing companies—all these contribute to a high level of communication to share knowledge of nematode control.

Nematode control will grow in importance. Better methods will result from global searches for genes that confer resistance to nematodes, for antinematode compounds in plants, for agents of biological control, and for cultural methods adapted to agriculture in diverse circumstances.

We all inhabit the same planet—we must cooperate in solving common problems, including management of nematodes.

REFERENCES

1. Alam, M. M., Khan, A. M., and Saxena, S. K. 1979. Mechanism of control of plant parasitic nematodes as a result of the application of organic amendments to the soil. V. Role of phenolic compounds. Indian J. Nematol. 9:136 – 142.

2. Barker, K. R. and Nusbaum, C. J. 1971. Diagnostic and advisory programs. In: Plant Parasitic Nematodes, Vol. 1 Morphology, Anatomy, Taxonomy, and Ecology. B. M. Zuckerman, W. F. Mai, and R. A. Rohde, eds., Academic Press, New York, pp. 281–301.

3. Egunjobe, O. A., and Olaitan, J. O. 1986. Response of *Meloidogyne incognita* infected cowpea to some agro-waste soil amendments. Nematropica 16:33 – 43.

4. Gommers, F. J., Bakker, J. and Wynberg, H. 1982. Dithiophenes as singlet oxygen sensitizers. Photochem. Photobiol. 35:615 – 619.

5. Hafez, S. C. and Raski, D. J. 1981. Residue dynamics and persistence of aldicarb and its biologically similar active metabolites in grapevines. J. Nematol. 13:29 – 36.

6. Hardison, J. R. 1980. Role of fire for disease control in grass seed production. Plant Dis. 64:641 – 645.

7. Jones, R. L. and Back, R. C. 1984. Monitoring aldicarb residues in Florida soil and water. Environ. Toxicol. Chem. 3:9 – 20.

8. Kahn, R. P. 1982. The host as a vector: exclusion as a control. In: Pathogens, vectors and plant disease: approaches to control. K. F. Harris and K. Maramorosch, eds. Academic Press, New York, pp. 123–149.

9. Katan, J. 1980. Solar pasteurization of soils for disease control: Status and prospects. Plant Dis. 64:450 – 454.

10. Katan, J., Greenberger, A., Alon, H., and Grinstein, A. 1976. Solar heating by polyethylene mulching for the control of diseases caused by soil-borne pathogens. Phytopathology 66:683 – 688.

11. Kerry, B. R. 1978. Natural control of the cereal cyst- nematode by parasitic fungi. ARC Res. Rev. (U.K.) 4:17 – 21.

12. Kerry, B. R. 1981. Fungal parasites: a weapon against cyst nematodes. Plant Dis. 65:390 – 393.

13. Kuiper, K. 1977. Introductie en Vestiging van planteparasitaire Aaltjes in nieuwe Polders in het bijzonder van *Trichodorus teres*. Meded. Landbouwhogesch. Wageningen. 77–4:1 – 139.

14. LaMondia, J. A. and Brodie, B. B. 1984. Control of *Globodera rostochiensis* by solar heat. Plant Dis. 68:474 – 476.

15. Mankau, R. 1981. Microbial Control of Nematodes. In: Plant Parasitic Nematodes, Vol. 3 (B. M. Zuckerman and R. A. Rohde, eds.). Academic Press, New York, Chapter 19, pp. 475–494.

16. Murphy, W. S., Brodie, B. B., and Good, J. M. 1974. Population dynamics of plant nematodes in cultivated soil: Effects of combinations of cropping systems and nematicides. J. Nematol. 6:103 – 107.

17. Rodriguez-Kabana, R., Morgan-Jones, G., and B. Ownley Gintis. 1984. Effects of chitin amendments to soil on *Heterodera glycines*, microbial populations, and colonization of cysts by fungi. Nematropica 14:10 – 25.

18. Rodriguez-Kabana, R., King, P. S. and Pope, M. H. 1981. Comparison of in-furrow applications and banded treatments for control of *Meloidogyne arenaria* in peanuts and soybeans. Nematropica 11:53 – 67.

19. Sayre, R. M. 1971. Biotic influences in soil environment. In: Plant Parasitic Nematodes, Vol. 1. (B. M. Zuckerman, W. F. Mai, and R. A. Rohde, eds.). Academic Press, New York, Chapter 9, pp. 235–256.

20. Southey, J.F. 1971. Physical methods of control. In: Plant Nematology (J. F. Southey, ed.). Ministry of Agriculture, Fisheries and Food, GD1, H. M. Stationery Office, London, Chapter 15, pp. 302–312.

21. Van Gundy, S. D. and McElroy, F. D. 1969. Sheath nematode, its biology and control. Proc. Int. Citrus Symp., 1st., Vol. 2:985 – 989; see also Van Gundy, S. D., McElroy, F. D., Cooper, A. F., and Stolzy, L. H. 1968. Influence of soil temperature, irrigation and aeration on *Hemicycliophora arenaria*. Soil Sci. 106:270 – 274.

22. Wright, D. J. 1981. Nematicides: Mode of action and new approaches to chemical control.In: Plant Parasitic Nematodes, Vol. 3 (B. M. Zuckerman and R. A. Rohde, eds.). Academic Press, New York, pp. 421- 449.

Author Index

Subject Index